MAINTENANCE OF MICROORGANISMS
AND CULTURED CELLS

Maintenance of Microorganisms and Cultured Cells

A Manual of Laboratory Methods

Second Edition

Edited by

B. E. KIRSOP

Microbial Strain Data Network
Institute of Biotechnology
Cambridge University
Cambridge, UK

and

A. DOYLE

European Collection of Animal Cell Cultures
PHLS Centre for Applied Microbiology Research
Porton Down
Salisbury, UK

Academic Press
Harcourt Brace Jovanovich Publishers
London   San Diego   New York   Boston
Sydney   Tokyo   Toronto

This book is printed on acid-free paper

ACADEMIC PRESS LIMITED
24–28 Oval Road
LONDON NW1 7DX

United States Edition published by
ACADEMIC PRESS INC.
San Diego, CA 92101

First edition published 1984

British Library Cataloguing in Publication Data
is available
ISBN 012–410351–0

Typeset by J&L Composition Ltd, Filey, North Yorkshire
Printed in Great Britain by Galliard (Printers) Ltd, Great Yarmouth

CONTENTS

Contributors vii

Preface ix

1. **Introduction** 1

2. **Service Collections: their Functions** 5
   B. E. Kirsop

3. **General Introduction to Maintenance
   Methods** 21
   J. J. S. Snell

BACTERIA

4.1. **Maintenance of Bacteria by
   Freeze-drying** 31
   R. H. Rudge

4.2. **Maintenance of Bacteria on Glass Beads
   at −60°C to −76°C** 45
   D. Jones, P. A. Pell and P. H. A. Sneath

4.3. **Maintenance of Bacteria in Gelatin Discs** . 51
   J. J. S. Snell and M. Kocur

4.4. **Maintenance of Industrial & Marine
   Bacteria & Bacteriophages** 57
   T. R. Dando and I. J. Bousfield

4.5. **Low-Temperature Freezing of
   Microorganisms on Silica Gel** 65
   T. M. Sidyakina

4.6. **Maintenance of Anaerobic Bacteria** 71
   C. S. Impey and B. A. Phillips

4.7. **Maintenance of Phototrophic Bacteria** . . . 81
   K. A. Malik

4.8. **Maintenance of Methanogenic Bacteria** . 101
   H. Hippe

4.9. **Maintenance of Leptospira** 115
   S. A. Waitkins

4.10. **Maintenance of Microorganisms by
   Simple Methods** 121
   K. A. Malik

FILAMENTOUS FUNGI

5.   **Maintenance of Filamentous Fungi** 133
      D. Smith

YEASTS

6.   **Maintenance of Yeasts** 161
      B. E. Kirsop

ALGAE AND PROTOZOA

7.1. **Maintenance of Algae and Protozoa** 183
      M. R. McLellan, A. J. Cowling, M. Turner
         and J. G. Day

7.2. **Maintenance of Parasitic Protozoa by
         Cryopreservation** 209
      E. R. James

ANIMAL CELLS

8.   **Maintenance of Animal Cells** 227
      A. Doyle and C. B. Morris

PLANT TISSUE CULTURES

9.   **Maintenance of Plant Tissue Cultures** . . . 243
      L. A. Withers

      References 269

      Appendix I: Documentation 287

      Appendix II: List of Suppliers 289

      Genus Index 293

      Subject Index 299

CONTRIBUTORS

I. J. BOUSFIELD, National Collection of Industrial and Marine Bacteria Ltd, 23 St Machar Drive, Aberdeen AB2 1RY, UK

A. J. COWLING, Culture Collection of Algae and Protozoa, Institute of Freshwater Ecology, The Windermere Laboratory, Far Sawrey, Ambleside, Cumbria LA22 0LP, UK

T. R. DANDO, National Collection of Industrial and Marine Bacteria Ltd, 23 St Machar Drive, Aberdeen AB2 1RY, UK

J. G. DAY, Culture Collection of Algae and Protozoa, Institute of Freshwater Ecology, The Windermere Laboratory, Far Sawrey, Ambleside, Cumbria LA22 0LP, UK

A. DOYLE, European Collection of Animal Cell Cultures, PHLS Centre for Applied Microbiology and Research, Division of Biologics, Porton Down, Salisbury, Wiltshire SP4 0JG, UK

H. HIPPE, DSM – Deutsche Sammlung von Mikroorganismen und Zellkulturen GmbH, Mascheroder Weg 1b, D–3300 Braunschweig, Germany

C. S. IMPEY, Cherry Valley Farms Ltd, North Kelsey Moor, Lincoln LN7 6HH, UK

E. R. JAMES, Medical University of South Carolina, Charleston, SC, USA

D. JONES, Department of Microbiology, School of Medical Science, University of Leicester, University Road, Leicester LE1 7RH, UK

B. E. KIRSOP, Microbial Strain Data Network, Institute of Biotechnology, Cambridge University, 307 Huntingdon Road, Cambridge CB3 0JX, UK

M. KOCUR, Czechoslovak Collection of Microorganisms, Masaryk University, Jostova 10, Brno 66243, Czechoslovakia

K. A. MALIK,   DSM–Deutsche Sammlung von Mikroorganismen und Zellkulturen GmbH, Mascheroder Weg 1b, D–3300 Braunschweig, Germany

M. R. McLELLAN,   Cell Systems Ltd, Orwell House, Cowley Road, Cambridge CB4 4WY, UK

C. B. MORRIS,   European Collection of Animal Cell Cultures, PHLS Centre for Applied Microbiology and Research, Division of Biologics, Porton Down, Salisbury, Wiltshire SP4 0JG, UK

P. A. PELL,   Department of Microbiology, School of Medical Science, University of Leicester, University Road, Leicester LE1 7RH, UK

B. A. PHILLIPS,   National Collection of Food Bacteria, AFRC Institute of Food Research, Reading RG2 9AT, UK

R. H. RUDGE,   National Collection of Type Cultures, Central Public Health Laboratory, 61 Colindale Avenue, London NW9 5HT, UK

T. M. SIDYAKINA,   All Union Collection of Microorganisms, Institute of Biochemistry and Physiology of Microorganisms, USSR Academy of Sciences, Pushchino, Moscow Region 142292, USSR

D. SMITH,   Culture Collection, International Mycological Institute, Ferry Lane, Kew, Surrey TW9 3AF, UK

P. H. A. SNEATH,   Department of Microbiology, School of Medical Science, University of Leicester, University Road, Leicester LE1 7RH, UK

J. J. S. SNELL,   Division of Microbiological Reagents and Quality Control, Central Public Health Laboratory, 61 Colindale Avenue, London NW9 5HT, UK

M. TURNER,   Culture Collection of Algae and Protozoa, Dunstaffnage Marine Laboratory, PO Box 3, Oban, Argyll, PA34 4AD, UK

S. A. WAITKINS,   Pathvet Services, Unit 16, Rotherwas Industrial Estate, Hereford, UK

L. A. WITHERS,   International Board for Plant Genetic Resources, c/o FAO, Via della Sette Chiese 142, 00145 Rome, Italy

PREFACE TO THE SECOND EDITION

Microbiology and cell biology have always depended upon the maintenance of biological material to provide viable and stable cultures. The increasing use of organisms for industrial environmental purposes reinforces this need and makes effective culture maintenance an essential prerequisite.

The first edition of *Maintenance of Microorganisms* filled a gap in the scientific literature by bringing together in one volume methods that had been developed and used successfully by those experienced in culture maintenance across a broad spectrum of biological material. It included chapters on bacteria (of general, medical and industrial importance; anaerobes, methanogens and leptospira), fungi (including yeasts), algae and protozoa.

As predicted in the first edition, the second edition up-dates preservation methodology in the light of new research and developments and also expands the coverage. Most chapters have been substantially revised; new sections have been added to include contributions on the preservation of further groups of sensitive bacteria and on the use of a number of simple methods that can be carried out with inexpensive and readily available equipment. Sections on the maintenance of plasmid-bearing strains of bacteria and yeasts have been added, with particular attention being paid to genetic stability. New chapters on animal cells and plant tissue cultures, which are increasingly important in research and industrial manufacturing processes, have been included. To encompass this new material, the title of the book has been amended to *Maintenance of Microorganisms and Cultured Cells*.

As in the first edition, some repetition has been deliberately accepted in the interests of ease of use, but with increased coverage it is now even more important to recognize that successful techniques applied to one group of microorganism or cell type may well be applicable to other groups. Although cross-referencing has been made to other relevent sections, readers are advised not to confine their attention only to the contributions relating to the organisms of interest. Indeed, added value may well be gained from the cross-fertilization of ideas between specialist microbiologists that a compilation such as this allows.

As before, the volume remains a practical guide. In some chapters, where technology is still experimental and precise methodology for a wide range of species cannot yet be recommended, more background discussion is included. The chapters do not, therefore, all conform to an exact format and the editors believe that, in the present state of the art, flexibility of this kind is necessary.

We thank the contributors who have written and revised the material and who through their willing support have made the compilation of this volume a pleasure.

B. E. Kirsop
A. Doyle

1. INTRODUCTION

B. E. KIRSOP and A. DOYLE[1]

Microbial Strain Data Network, Institute of Biotechnology,
Cambridge University, Cambridge, UK

[1] European Collection of Animal Cell Cultures,
PHLS Centre for Applied Microbiology and Research,
Porton Down, Salisbury, UK

The ability to preserve successfully a wide range of microorganisms and cell cultures has been a major achievement in biology over the last century that many have taken for granted. The need to conserve biological material such as blood cells, animal semen and starter cultures provided a strong stimulus to research, and the basic principles affecting cell survival and cell death have come nearer to being understood at the molecular level. As a result, culture preservation has reached a high level of success at a time when the needs of biotechnologists and environmental biologists have become more demanding.

Although a number of excellent preservation techniques have now been developed, they must be correctly applied if contamination, loss of viability and genetic change are to be prevented. Reliable stock cultures are needed for a variety of reasons. Teaching laboratories need a library of cultures that exhibit typical reactions; industrial laboratories need to maintain production strains, in addition to a bank of biological material for screening purposes; medical laboratories need reference strains both for routine testing in pathology laboratories and for research; taxonomists must maintain large numbers of cultures for comparative studies; and research laboratories need pure cultures for a wide range of purposes.

Despite the fundamental importance of a reliable supply of pure and stable cultures, culture preservation is often afforded low priority and is often carried out with inadequate staffing levels and inferior equipment. The importance of culture preservation and the value of investing in staff and equipment for this purpose is often only recognized when valuable or unique

MAINTENANCE OF MICROORGANISMS 2nd Edn
ISBN 012–410351–0

strains are lost. Industrial, research or teaching collections are often housed in laboratories where culture maintenance is not a primary function and little time is available to develop or assess different preservation methods. The service culture collections, however, have a full commitment to the preservation of cultures and their accumulated expertise is of great value to others.

Information on preservation methods is thinly distributed through the literature and often not readily available; this manual assembles in one volume a collection of reliable preservation methods for a diverse range of organisms. Guidance is provided on the selection of the most appropriate method for particular circumstances. Technical details of methods are presented, and their reliability with regard to expected survival levels, shelf-life and stability is assessed. The relative costs of different methods, sources of equipment and materials, and the relevant literature references are given.

It must be stressed that there is no universal method for the successful preservation of all microorganisms and cell cultures. Different taxonomic groups, and even strains within a species, exhibit significant differences in response to the stresses imposed by culture preservation and resuscitation. The methods described in this manual, when used as described and for the defined group of organisms, will generally prove successful, but occasions will arise when further investigation will be needed to achieve complete success.

The manual is practical in nature and aims to provide details which the authors have found to work well. Each chapter covers a number of methods that have been found successful for a particular group of microorganisms or cell cultures. These methods may, however, prove successful for other cell types, and this should be borne in mind when using the book.

Repetition of detail occurs in a number of chapters, but since the book will be used primarily as a practical manual these have been deliberately retained to avoid the need for repeated reference to other sections to obtain essential methodology. The names and addresses of a number of suppliers of equipment quoted in the text are supplied in Appendix II. Although this list is not exhaustive, and is uneven in coverage of different geographical regions, it nevertheless provides a lead to possible suppliers. Further information can also be obtained by contacting any of the service collections listed in Chapter 2.

The support available to the scientific community from the service collections is not always recognized, and Chapter 2 provides a description of the role they play in microbiology. Further information on services in Europe can be obtained from the European Culture Collection Information Centre; world-wide information is available from the World Data Center on

Collections of Microorganisms or the Microbial Strain Data Network. Contact information on these centres is listed in Chapter 2.

The methods described in this volume provide reliable information on how to preserve cultures successfully; nevertheless, continued monitoring of viability levels, genetic stability and purity must be carried out for the cultures to serve as completely reliable inocula for experimental, industrial or educational purposes. Guidance is given in Appendix I on documentation that could be adopted for maintaining proper records.

It is the aim of this manual to raise the quality of the biological material provided by collections and, in turn, the level of scientific activity. It is hoped that the first edition has gone some way towards achieving this aim and that the additional information contained in this second edition will continue the process.

2. SERVICE COLLECTIONS: THEIR FUNCTIONS

B. E. KIRSOP

Microbial Strain Data Network, Institute of Biotechnology,
Cambridge University, Cambridge, UK

The main function of service culture collections is to act as depositories for all kinds of microorganisms that are of past, present or potential importance, so that resource and information centres for the general support of microbiology are established. The basic function of culture collections, therefore, is the collection, maintenance and supply of cultures.

The organisms collected reflect the interests of the collection and are obtained in one of several ways. Following searches through the scientific literature, direct application may be made to authors to deposit strains; alternatively, microbiologists may themselves make approaches to appropriate collections. Again, collections themselves may deposit cultures that have formed part of their own taxonomic or identification activities. Deposited cultures are checked for purity and authenticity before accession by the collection.

The selection of maintenance methods that produce maximum survival levels and strain stability is of fundamental importance, since strain drift in stocks maintained in service collections is unacceptable. Assessments of survival are made immediately after processing and at intervals during storage, so that fall-off in viability is monitored and an effective maintenance programme established. In order to assure a reliable service for the supply of cultures, quality control measures are carried out routinely. In addition, appropriate stock-holding levels are determined and administrative arrangements are established for handling orders and invoicing customers; postal regulations are followed for the despatch of cultures at home and overseas.

These activities, common to all service culture collections, generate a quantity of essential information that must be systematically recorded. Many collections are now using computers for this purpose. They are ideally suited to the storage and management of both customer records and stock supplies, and

MAINTENANCE OF MICROORGANISMS 2nd Edn
ISBN 012–410351–0

greatly streamline the business administration of collections. Again, after appropriate coding, strain data is generally stored in a computer, allowing rapid searching and retrieval of scientific information, and establishing a database that may be used subsequently for computer identification. Computers are increasingly being used for the storage of literature references and are equally appropriate for the preparation of catalogues. An up-to-date catalogue is an essential requirement for the effective functioning of a service collection and, in the past, its preparation was an onerous and lengthy job. The transfer of catalogue data to a computer, however, allows ready updating for future editions and provides copy that can be printed directly from disk or tape, thus eliminating the need for conventional proof-reading.

In addition to the three basic activities (the collection, maintenance and supply of cultures, and the record-keeping associated with them), culture collections provide a number of other services. As a result of the expertise developed by collection staff and the close contacts established with microbiologists using their strains, collections become information centres for all matters relating to the microorganisms they hold. Many collections are now collaborating in the development of on-line databases and networks (see Table I) in order to make the information they hold readily available to the scientific community. In addition to supplying answers to written and telephoned enquiries, staff are called upon to lecture, help on courses and contribute to scientific journals and books. Much of the expertise existing in the collections is of a taxonomic nature and most service collections are able to provide an identification service. This is particularly valuable for microbiologists, since identification may be lengthy, expensive, demand sophisticated techniques, and yet only be required intermittently.

A further service of particular interest to industry is the provision of 'safe-deposit' facilities. Several collections are able to maintain important strains under optimum conditions, while providing access only to the depositor. This enables industry to depend on a reliable back-up for the preservation of strains that may be of crucial importance to production. Again, collections may provide a culture preservation service in which strains are expertly preserved and returned to the customer for storage. This may be useful both to industry, which is assured of a reliable supply of a 'standard inoculum' for production purpose, or to research in which certain experiments may require the use of aliquots of a single population of cells.

With the present rapid developments in biotechnology there is an increasing need to patent processes or, more recently, genetically engineered strains, and there is a need for recognized

Table I: Information centres.

Regional

ICECC — for all information on the services provided by the European Culture Collections

Information Center for European Culture Collections
Mascheroder Weg 1
4400 — Braunschweig
Germany

Telephone: +39-531-618715
Fax: +39-531-618718
Telex: 531104
Electronic mail: Telecom Gold 75: DBI0274

International

MSDN — for international communications and links to databases related to microbiology and biotechnology

Microbial Strain Data Network
Institute of Biotechnology
Cambridge University
307 Huntingdon Road
Cambridge CB3 0JX, UK

Telephone: +44-223-276622
Fax: +44-223-277605
Telex: 81240 Camspl G
Electronic mail: Telecom Gold 75: DBI0001

WDC — for a Directory of Culture Collections worldwide and other associated databases

World Data Center for Collections of Cultures of Microorganisms
RIKEN
2–1 Hirosawa, Wako
Saitama 351–01
Japan

Telephone: +81-484-62-1111
Fax: +81-484-64-5651
Electronic Mail: BT TYMNET 42: CDT0007

depositories. Many culture collections have now become accepted as International Depository Authorities and are able to provide the legal requirements for the purpose. Acceptance of collections for this role demands certain assurances regarding permanence, impartiality and the existence of appropriate scientific expertise and facilities.

It is clear from a consideration of these varied activities that culture collections are not merely collections of cultures but provide a number of essential services for the general support of biotechnology and environmental microbiology. In addition, it has always been considered important for collections to carry out research programmes that complement their interests. This not only ensures awareness of developments in their own fields but takes the maximum scientific advantage of having large numbers of authenticated microorganisms available for study.

Clearly, one area of research of particular interest to culture collection laboratories is taxonomy and its relevance to identification procedures. Again, preservation methods (particularly cryopreservation) and problems related to cell survival and strain stability are of fundamental importance. There is, moreover, a growing awareness of the extent to which many of the microorganisms maintained in the culture collections have not been fully characterized, and collections have a developing interest in screening procedures to enable the research into and industrial potential of the cultures maintained to be fully explored. The diversity of the types of microorganisms maintained, the interest of parent institutes and the changing requirements of microbiology lead to substantial variety in the kinds of research undertaken by collections.

I. EUROPEAN CULTURE COLLECTIONS

Chapters in this volume have been contributed for the most part by curators of service culture collections in Europe. Many of these centres are members of the European Culture Collections' Organization (ECCO). Their names, addresses and other contact information are provided in Tables II and III. Further information on the activities of European Culture Collections can be obtained from the European Information Centre for Culture Collections (ICECC; see Table I) which has been established with support from the Commission of the European Community and ECCO.

Within ECCO membership there exists a great deal of experience in the practical aspects of culture preservation, and also an extensive wealth of knowledge on the wider aspects of culture 'maintenance' that embraces accessioning procedures, record-keeping, computer databases, distribution and safety regulations, as well as the essential underpinning knowledge of taxonomy and

Table II: Culture collections in Europe that are members of the European Culture Collections' Organization (ECCO).

Culture collection (Listed alphabetically by country)		Acronym of collection	Groups of organisms held
Belgium			
Dr N. Nolard-Tintigner INSTITUTE OF HYGIENE AND EPIDEMIOLOGY Mycology Laboratory Rue Juliette Wytsman, 14 B-1050 BRUSSELS	Tel: +32-2-6425630 Fax: +32-2-6425519 Telex: 21034 ihebru BT TYMNET/Telecom Gold: 75: DBI0248	IHEM	Fungi
Dr K. Kersters LABORATORIUM VOOR MICROBIOLOGIE EN MICROBIELE GENETICA Faculteit der Wetenschappen K.L. Ledeganckstraat 35 B-9000 GENT	Tel: +32-91-227821 ext. 435 or 434 Fax: +32-91-223149 Telex: 12754 rug EARN gels@lmgmicro. rug.ac.be. BT TYMNET/Telecom Gold: 75: DBI0249	LMG	Bacteria
Prof. Dr G. L. Hennebert MYCOTÈQUE DE L' UNIVERSITÉ CATHOLIQUE DE LOUVAIN Place Croix du Sud 3, Bte 8 B-1348 LOUVAIN-LA-NEUVE	Tel: +32-10-473742 Fax: +32-10-451501 Telex: 59037 Ucl b BT TYMNET/Telecom Gold: 75: DBI0250	MUCL	Fungi, yeasts
Bulgaria			
Dr I. Tepavitcharova NATIONAL BANK FOR INDUSTRIAL MICROORGANISMS AND CELLULAR CULTURES 25 'Lenin' Bd, Bl 2 1113 SOFIA	Tel: +359-2-720865 Telex: 23468	NBIMCC	Animal cells, bacteria, fungi, plasmids, yeasts
Czechoslovakia			
Dr L. Valicek (B/3), VIRUSES COLLECTION OF ANIMAL PATHOGENIC MICROORGANISMS Veterinary Research Institute CS-621 32 BRNO	Tel: +42-5-741121 Telex: 62475	CAPM	Viruses

Table II. *Continued.*

Culture collection (Listed alphabetically by country)		Acronym of collection	Groups of organisms held
Dr M. Vanova CULTURE COLLECTION OF FUNGI Department of Botany Faculty of Science Charles University Benatska 2 CS-128 01 PRAGUE 2	Tel: +42-2-297941	CCF	Fungi
Dr M. Kocur CZECHOSLOVAK COLLECTION OF MICROORGANISMS Masaryk University Jostova 10 662 43 BRNO	Tel: +42-5-23407	CCM	Bacteria, fungi
Ing. R. Kovacovska CZECHOSLOVAK COLLECTION OF YEASTS Slovak Academy of Sciences Centre of Chemical Research Institute of Chemistry Dubravska Cesta 9 CS-842 38 BRATISLAVA	Tel: +42-7-3782625	CCY	Yeasts
Finland			
Dr Maija-Liisa Suihko VTT COLLECTION OF INDUSTRIAL MICROORGANISMS VTT Biotechnical Laboratory Finland Tietotie 2 SF-02150 ESPOO	Tel:+358-0-4565133 Fax: +358-0-4552028 Telex: 122972 vttha sf EARN nsuiko@uk.ac. afrc.frin.	VTT	Bacteria, fungi, yeasts

Table II. *Continued.*

Culture collection (Listed alphabetically by country)		Acronym of collection	Groups of organisms held
France			
Dr L. Gardan COLLECTION FRANCAISE DES BACTERIES PHYTOPATHOGENES I.N.R.A. Station de Pathologie Vegetale et Phytobacteriologie Route de Saint-Clement Beaucouze F-49000 ANGERS	Tel: +33-41735161 Fax: +33-41735101 Telex: inrager 720565 f	CFBP	Bacteria
Mme Y. Cerisier COLLECTION NATIONALE DE CULTURES DE MICROORGANISMES Institut Pasteur 25, rue du Docteur Roux F-75724 PARIS Cedex 15	Tel: +33-1-45688251 Fax: +33-1-45688236 Telex: 250609 pasteur f	CNCM	Bacteria, fungi
Dr M. F. Roquebert MUSEUM NATIONALE D'HISTOIRE NATURELLE Laboratoire de Cryptogamie 12, rue Buffon F-75005 PARIS	Tel: +33-1-40793194 Fax: +33-1-40793484 Telex: musnahn 202641	LCP	Fungi
Germany			
Dr H. Prauser ZENTRALINSTITUT FÜR MIKROBIOLOGIE UND EXPERIMENTELLE THERAPIE Beutenbergstrasse 13 6900 JENA FRG	Tel: +37-78-852433 Fax: +37-78-31325 Telex: 588642 zimt dd	IMET	Bacteria

Table II. *Continued.*

Culture collection (Listed alphabetically by country)		Acronym of collection	Groups of organisms held
Dr D. Claus DSM-DEUTSCHE SAMMLUNG VON MIKRO-ORGANISMEN UND ZELLKULTUREN GmbH Mascheroder Weg 1 b D-3300 BRAUNSCHWEIG FRG	Tel: +49-531-61870 Fax: +49-531-618718 Telex: Teletex: 531104= DSM DFN (EAN): dsm@venus.gbf-braunschweig.dbp.de BT TYMNET/Telecom Gold: 75: DBI0225	DSM	Bacteria, plasmids, fungi, yeasts plant cells, animal cells

Hungary

Dr B. Lanyi, Dr M. Konkoly-Thege HUNGARIAN NATIONAL COLLECTION OF MEDICAL BACTERIA National Institute of Hygiene Gyali ut 2–6 H-1097 BUDAPEST	Tel: +36-1-142250 Telex: 225349 oki h	HNCB	Bacteria
Prof. Dr T. Deak NATIONAL COLLECTION OF AGRICULTURAL INDUSTRIAL MICROORGANISMS University of Horticulture and Food Industry Department of Microbiology and Biotechnology Somloi ut 14–16 H-1118 BUDAPEST	Tel: +36-1-665411 Fax: +36-1-666220 Telex: 226011 uhort Telecom Gold: 75: DBI0185	NCAIM	Bacteria, yeasts

Italy

Prof. A. Martini COLLECTIONE LIEVITI INDUSTRIALI Dipartimento di Biologia Vegetale Universita di Perugia 74 Borgo 20 Giugno 06100 PERUGIA, I	Tel: +39-75-30458 Fax: +39-75-31766 Telex: unipg i 662078	DBVPG	Yeasts

Table II. *Continued.*

Culture collection (Listed alphabetically by country)		Acronym of collection	Groups of organisms held
The Netherlands			
Dr J. A. Stalpers CENTRAALBUREAU VOOR SCHIMMELCULTURES P.O. Box 273 Oosterstraat 1 NL-3740 AG BAARN	Tel: +31-2154-81211 Fax: +31-2154-16142 BT TYMNET/Telecom Gold: 75: DBI0548	CBS	Fungi, actino- mycetes, yeasts
Norway			
Dr O. M. Skulberg CULTURE COLLECTION OF ALGAE Norwegian Institute for Water Research P.O. Box 69 Korsvoll N-0808 OSLO 8	Tel: +47-2-235280 Fax: +47-2-394189 Telex: 74190 niva n	NIVA	Algae
Poland			
Prof. Dr M. Modarski, Dr Rowinski POLISH COLLECTION OF MICROORGANISMS Institute of Immunology and Experimental Therapy Polish Academy of Sciences Czerska 12 PL-53 114 WROCLAW POLAND		PCM	Bacteria
Portugal			
Prof. I. Spencer-Martins PORTUGUESE YEAST CULTURE COLLECTION Gulbenkian Institute of Science 2781 Oeiras Codex PORTUGAL	Tel: +351-1-4431436 Fax: +351-1-4431631 BT TYMNET/Telecom Gold: 75: DBI0261	IGC	Yeasts

Table II. *Continued.*

Culture collection (Listed alphabetically by country)		Acronym of collection	Groups of organisms held

Spain

| Prof Dr F. Uruburu COLECCION ESPANOLA DE CULTIVOS TIPO Departamento de Microbiologia Facultad de Ciencias Biologicas Universidad de Valencia E-46100 BURJASOT | Tel: +34-6-3864300 or 3864384 Fax: +34-6-3864372 BT TYMNET: 75:CDT0428 | CECT | Bacteria, fungi, yeasts |

Sweden

| Civ. Ing. E. Falsen CULTURE COLLECTION, UNIVERSITY OF GOETEBORG Dept. of Clinical Bacteriology Guldhedsg 10 S-413 46 GOETEBORG | Tel: +46-31-604625 Fax: +46-31-825484 Telex: BT GOLD (UK) 265451 or 265871 in first line of text put Ref: DBI0070 EARN node SEGUC21 Userid: TIXCC BT TYMNET/Telecom Gold: 75: DBI0070 | CCUG | Bacteria, yeasts |
| Dr O. Constantinescu UPPSALA UNIVERSITY CULTURE COLLECTION OF FUNGI The Herbarium (Fytoteket) University of Uppsala P.O. Box 541 S-751 21 UPPSALA | Tel: +46-18-182794 FYTOOCSEMAX51.BITNET | UPSC | Fungi |

Switzerland

| Dr M. R. Darekar CENTRE DE COLLECTION DE TYPE MICROBIEN Institut de Microbiologie Université de Lausanne 44 Rue de Bugnon CH-1011 LAUSANNE | Tel: +41-21-412339 Fax: +41-21-3143181 Telex: 25012 | CCTM | Bacteria |

Table II. *Continued.*

Culture collection (Listed alphabetically by country)		Acronym of collection	Groups of organisms held
Turkey			
Dr Emel Tumbay Department of Microbiology Faculty of Medicine Ege University BORNOVA, IZMIR			Fungi
Prof. Dr E. Tali Cetin CENTRE FOR RESEARCH AND APPLICATION OF CULTURE COLLECTIONS OF MICROORGANISMS Istanbul Faculty of Medicine Microbiology Department Temel Bilimler Binasi Capa-Topkapi ISTANBUL	Tel: +90-1-5255504 or 5250904	KUKENS	Bacteria
United Kingdom			
M. F. Turner CULTURE COLLECTION OF ALGAE AND PROTOZOA Dunstaffnage Marine Laboratory PO. Box 3 OBAN Argyll PA34 4AD	Tel: +44-631-62244 Fax: +44-631-65518 Telex: 776215 – MARLAB G JANET: S-CCAP @ UK.AC. NSM.VA	CCAP	Marine algae
Dr J. Day CULTURE COLLECTION OF ALGAE AND PROTOZOA Institute of Freshwater Ecology The Ferry House Far Sawrey Ambleside Cumbria LA22 0LP	Tel: +44-9662-2468/9 Fax: +44-9662-6914 Telex: 94070416 — WIND G JANET: WI-CCAP @ UK. AC.NWI.VA	CCAP	Algae, protozoa

Table II. *Continued.*

Culture collection (Listed alphabetically by country)		Acronym of collection	Groups of organisms held
Dr D. Smith INTERNATIONAL MYCOLOGICAL INSTITUTE Culture Collection & Industrial Service Division Ferry Lane, KEW Surrey TW9 3AF	Tel: +44-71-9404086 Fax: +44-71-3321171 Telex: 9312102252 mi g; BT Gold UK 265871 (monref g) and quote CAU009 BT TYMNET/Telecom Gold: 75: DBI0019	IMI	Fungi
Dr. A. Doyle EUROPEAN COLLECTION OF ANIMAL CELL CULTURES PHLS Centre for Applied Microbiology and Research Porton Down SALISBURY Wiltshire SP4 0JG	Tel: +44-980-610391 Fax: +44-980-611096 Telex: 47683 phcamr g BT TYMNET/Telecom Gold: 75: DBI0008	ECACC	Animal cells
Dr P. Jackman NATIONAL COLLECTION OF FOOD BACTERIA AFRC Institute of Food Research Reading Laboratory Shinfield READING RG2 9AT	Tel: +44-734-883103 Fax: +44-734-884763 Telex: 9312102022 JANET: phillips@uk.ac. afrc.frin BT TYMNET/Telecom Gold: 75: DBI0013	NCFB (NCDO)	Bacteria
Dr. I. Bousfield NATIONAL COLLECTIONS OF INDUSTRIAL AND MARINE BACTERIA Ltd. 23 St. Machar Drive ABERDEEN AB2 1RY	Tel: +44-224-273332 Fax: +44-224-487658 Telex: 73458 uniabn g BT TYMNET/Telecom Gold: 75: DBI0271	(NCIB, NCMB)	Bacteria
Dr C. K. Campbell NATIONAL COLLECTION OF PATHOGENIC FUNGI PHLS Mycological Reference Laboratory Central Public Health Laboratory 61 Colindale Avenue LONDON NW9 5HT	Tel: +44-81-2004400 Fax: +44-18-2007874 Telex: 8953942 defend g	NCPF	Fungi

Table II. *Continued.*

Culture collection (Listed alphabetically by country)		Acronym of collection	Groups of organisms held
Dr D. E. Stead NATIONAL COLLECTION OF PLANT PATHOGENIC BACTERIA Central Science Laboratory Hatching Green HARPENDEN Hertfordshire AL5 2BD	Tel: +44-5827-5241 Fax: +44-5827-62178 Telex: 826363	NCPPB	Bacteria
Dr L. R. Hill NATIONAL COLLECTION OF TYPE CULTURES PHLS Central Public Health Laboratory 61 Colindale Avenue LONDON NW9 5HT	Tel: +44-81-2004400 Fax: +44-81-2007874 Telex: 8953942 defend g BT TYMNET/Telecom Gold: 75: DBI0086	NCTC	Bacteria, myco- plasmas, bacterio- phages, plasmids
Dr J. Carey NATIONAL COLLECTION OF WOOD ROTTING FUNGI Biodeterioration Section Timber Division Building Research Establishment Garston WATFORD Hertfordshire WD2 7JR	Tel: +44-923-894040 Fax: +44-923-662184 Telex: 923220	NCWRF	Wood rotting fungi
Mr C. Bond (Acting Curator) NATIONAL COLLECTION OF YEAST CULTURES AFRC Institute of Food Research Norwich Laboratory Colney Lane NORWICH NR4 7UA	Tel: +44-603-56122 Fax: +44-603-58939 Telex: 975452 firinor g JANET: jackman@uk.ac. afrc.frin BT TYMNET/Telecom Gold: 75: DBI0013	NCYC	Yeasts

Table II. *Continued.*

Culture collection (Listed alphabetically by country)		Acronym of collection	Groups of organisms held
USSR			
Prof. Dr L. V. Kalakoutskii ALL-UNION COLLECTION OF MICROORGANISMS Institute of Biochemistry and Physiology of Microorganisms USSR Academy of Sciences PUSCHINO Moscow Region 142292	Tel: +7-095-2316576 Telex: 412609 micro su	(BKM, VKM)	Bacteria, fungi, yeasts
Dr M. G. Vladimirova COLLECTION OF MICROALGAE OF THE INSTITUTE OF PLANT PHYSIOLOGY USSR Academy of Sciences Botanicheskaya 35 MOSCOW 127276	Tel: +7-095-4822904	IPPAS	Algae
Institute of Genetics and Selection of Industrial Microorganisms I Dorozhny proezd, 1 MOSCOW 113545	Tel: +7-095-3153774		
Yugoslavia			
Prof. Dr A. Cimerman CULTURE COLLECTION OF FUNGI Kemijski Institut 'Boris Kidric' Hajdrihova 19 YU-61115 LJUBLJANA	Tel: +38-61-263061 Fax: +38-61-263285 Telex: 32121 kibklj yu	MZKIBK	Fungi

culture identification. The members of ECCO meet once a year to exchange information and experiences in the different aspects of administering publicly available resource centres. The aim of the organization is to foster collaboration and to provide support to its member collections. Currently there are 42 members from 19 countries.

ECCO is an adhering member of the international World

Table III: Some of the major culture collections worldwide that publish catalogues*

American Type Culture Collection (ATCC)
12301 Parklawn Drive
RockvilledMaryland 20852
USA
Telephone: +1 301 231 5578
Electronic Mail: BT TYMNET 42:CDT0109

Japan Collection of Microorganisms (JCM)
RIKEN
Wako
Saitama 351–01
JAPAN
Telephone: +81 484 64 5651
Electronic Mail: via WDC, BT TYMNET 42:CDT0007

Culture Collection of the Institute for Fermentation (IFO)
Institute for Fermentation
17–85 Jugo-Honchmachi 2-chome
Yodogawa-ku
Osaka
JAPAN
Telephone: +81 6 302 7281

Center for General Microbiological Culture Collection
China Committee for Culture Collections of Microorganisms
Institute of Microbiology
Academia Sinica
Beijing
CHINA

Culture Collection of the Base de Dados Tropical
Fundaïao Tropical de Pesquisas e Tecnologia 'Andrèe Tosello'
Rua Latina Coelho, 1301
Caixa Postal 1889
13.085 Campinas
São Paulo
BRAZIL
Telephone 192 42 7827
Electronic mail: BT TYMNET 42:CDT0094

* The Directory of over 300 culture collections is available from the World Data Center for Collections of Cultures of Microorganisms at the same address as JCM above.

Federation for Culture Collections (WFCC), and as such promotes the aims of the Federation. The WFCC operates through a structure of committees covering such matters as education, patents, safety, standards, information, publicity and endangered collections. It is concerned with the conservation of the microbial gene pool worldwide, and with the development of new ways of isolating and preserving cultures and studying them taxonomically. The WFCC represents culture collections at meetings of regulatory and funding bodies and at international meetings addressing such global issues as biodiversity.

The service culture collections throughout the world provide information, advice and cultures on request, and the reader is recommended to contact them on matters that fall within their areas of competence. Similarly, both ECCO and the WFCC can provide assistance in many ways.

The contact addresses for ECCO are:

Acting Secretary:
ICECC, Mascheroder
Weg 1, 4400–Braunschweig
Germany

Professor L. Kalakoutskii,
Chairman
All-Union Collection of
Microorganisms
Institute of Biochemistry and
Physiology of Microorganisms
USSR Academy of Sciences
Moscow Region 142292, USSR

The contact addresses for the WFCC are:

Dr V. Canhos, Secretary
Tropical Data Base
Fundacao Tropical de
Pesquisas e Tecnologia
'Andree Tosello'
Rua Latino Coelho 1301
Caixa Postal 1889
13.085 Campinas
São Paulo, Brasil

Ms Barbara Kirsop, President
Microbial Strain Data Network
Institute of Biotechnology
Cambridge University
307 Huntingdon Road
Cambridge CB3 0JX, UK

3. GENERAL INTRODUCTION TO MAINTENANCE METHODS

J. J. S. SNELL

Division of Microbiological Reagents and Quality Control,
Central Public Health Laboratory, London, UK

I. Choice of maintenance method 21
   A. Maintenance of viability 22
   B. Population change through selection 22
   C. Genetic change 22
   D. Purity 22
   E. Expense 22
   F. Number of cultures 23
   G. Value of cultures 23
   H. Supply and transportation of cultures 23
   I. Frequency of use of cultures 23
II. Methods 24
   A. Subculture 24
   B. Drying 25
   C. Freeze-drying 27
   D. Freezing 29

I. CHOICE OF MAINTENANCE METHOD

A wide variety of techniques are available for the preservation of microorganisms and it may be difficult to choose the method most suitable for a particular need. This chapter provides a summary of the main features of various available methods. The reader is also directed to the reviews of maintenance methods by Lapage and Redway (1974) and Lapage *et al.* (1978). All methods have their unique advantages and disadvantages and the choice of method for a particular use should be determined by relating the features of each method to the needs of the user. Features to be considered are described below.

MAINTENANCE OF MICROORGANISMS 2nd Edn
ISBN 012–410351–0

A. Maintenance of viability

Cell death may occur during the preservation process and there may be further losses during storage, eventually resulting in unacceptably low levels of viability. To avoid loss of the culture it must be resubjected to the preservation process. The method used should minimize loss of viability during processing and storage so that, once preserved, cultures will survive for long periods.

B. Population change through selection

A proportion of the cells in a population may die during the initial preservation process and this may appear of little significance to the user if high initial cell concentrations are used. However, the reduction in the number of viable cells may result in the selection of a resistant population of surviving cells and so introduces the possibility of change in the characteristics of the preserved culture. The preservation method should therefore retain the greatest number of viable cells so that the surviving population resembles the original as closely as possible.

C. Genetic change

Microorganisms are usually preserved because certain strain characters are of scientific or industrial significance. It is important that preserved strains do not lose important characters or gain others. Changes may occur during preservation through mutation or loss of plasmids so the preservation method should minimize the occurrence of these events.

D. Purity

Cultures preserved for most applications should remain pure and the preservation method should minimize the chance of contamination.

E. Expense

The cost of maintaining cultures includes the costs of staffing, equipment, materials and general facilities such as storage space and power supplies. The high capital cost of equipment for some methods such as freeze-drying may be offset against reduced staffing costs, since the long-term stability of cultures reduces the need for frequent manual intervention.

F. Number of cultures

The main factor to be considered in relating the number of cultures to the choice of method is the amount of operator time required for initial preservation and subsequent manipulation. A method found suitable for preservation of a small collection may be too labour-intensive when the number of strains increases. Choice of method for larger numbers of cultures may be affected by the amount of storage space available.

G. Valuc of cultures

The consequences of loss of a culture should be considered in choosing a preservation method. Important cultures should be preserved by methods that minimize the risk of loss, and for complete security more than one method should be used. For less important cultures criteria such as cost may carry more weight.

H. Supply and transportation of cultures

If cultures are to be distributed, replicates of the culture are needed. These may be prepared as required or prepared in bulk and stored for later distribution. The convenience of either approach depends on the preservation method used and the number of cultures to be distributed. If cultures are to be supplied through the post they must be in a form suitable for packing and must survive the delays and conditions likely to be encountered. Strict adherence to national and international postal regulations should be made. Copies of these regulations may be obtained from national postal authorities.

I. Frequency of use of cultures

Some cultures, such as assay strains, industrial production strains or those used for quality control, may be used frequently within a laboratory. In these cases, ease of resuscitation and the risk of contamination of stock cultures need to be considered. Other cultures may be used rarely and other factors may be of more importance.

No single method of preservation fulfils all these criteria and selection of a method will be a process of compromise. It is important to recognize that microorganisms, even within a species, differ in their tolerance to various preservation methods and, unless a collection is very specialized, it is unlikely that a single method will provide optimum conditions for all strains. Again, the choice will depend on a balance of advantages and disadvantages.

Most of the methods described in this chapter are discussed in greater detail in other sections of the manual and their description here is limited to an outline of the general principles involved and an indication of how the methods relate to the criteria discussed above. Some methods have been applied only to restricted groups of microorganisms and it is not possible to comment on their general usefulness. However, it may well be that methods developed for a particular group may have wider applications, and it is hoped that this manual will encourage investigations into the suitability of methods for microorganisms other than those for which they were originally described.

II. METHODS

A. Subculture

This method consists of inoculation of a suitable medium contained in a tube or bottle, incubation at an appropriate temperature to obtain growth, and storage under suitable conditions. The process is repeated at intervals that ensure the preparation of a fresh culture before the old one dies. The time that may be allowed to elapse between subcultures without risk of losing the culture depends primarily on the particular microorganism. Thus many bacteria such as staphylococci and coliforms will survive for several years, whereas the more delicate *Neisseria* spp. may require subculture after only a few weeks.

This method is inexpensive in terms of equipment but may be labour-intensive if organisms that require frequent subculture are maintained. Cultures are easily resuscitated as a further subculture is all that is required to obtain an active culture. The method is applicable to a wide range of microorganisms. However, contamination is a major problem and this risk occurs with each subculture. Apart from the undesirability of mixed cultures, any contaminants may outgrow and kill the original culture. Contamination may be reduced by sound microbiological technique and by pre-incubation of media before use. A two-tube method, where one is kept as a seed stock and another as a working culture, reduces the risk of contamination of the stock through frequent manipulation.

The risk of mislabelling or transposition of cultures is high, particularly where strains need frequent subculture. The risk increases with operator fatigue and can be alleviated to some extent by placing the numbered containers in random order, thus maintaining the concentration of the operator. Labels should be printed or typed, as handwriting is open to misinterpretation and numbers may become completely altered over time.

Loss of viability is a constant hazard with this method. If strains with different survival characteristics are maintained an established protocol is essential to ensure timely subculture of all strains. A system for this purpose has been described by Skerman (1973). Loss of cultures is usually sporadic, but whole batches may be lost because of such factors as faulty media. Dehydration may occur through imperfections in the seal of the container. Screw caps are not entirely free from the problem and plastic caps in particular may fail to seal a high proportion of the containers.

Cultures preserved by this method are very prone to loss of stability of characters, and the risk of change increases with the frequency of subculture. A large inoculum reduces the risk of selection, but increases the risk of contamination. Neither is the method particularly convenient for the distribution of cultures, since a subculture must be prepared and checked for purity before despatch and postal conditions may adversely affect viability. The media used may affect survival time. Although a great variety of media have been used for the storage of various microorganisms there is a tendency to use unenriched media with limited nutrients. Storage in water has been found suitable for some organisms (Berger, 1970). For most microorganisms excess carbohydrates should not be included in the medium since the acid produced might kill the culture.

Storage periods can be extended by reducing the metabolic rate of the microorganisms. This may be achieved by restricting the availability of air to the culture; liquid paraffin, as a layer on the surface of the culture, has been used extensively for this purpose, especially with fungi (Onions, 1971). Metabolic activity may also be reduced by lowering the storage temperature, and storage at 5°C has been widely used for a variety of microorganisms. However, not all microorganisms survive longer in culture at lower temperatures; *Neisseria* spp., for example, appear to survive better at 37°C.

B. Drying

Desiccation as a means of preserving microorganisms has been extensively used. A wide variety of methods exist, all consisting essentially of removal of water and prevention of rehydration. The methods have been most widely applied to fungi, which appear more resistant to drying than other groups of microorganisms. Some yeasts and selected bacteria have also been successfully preserved by drying.

1. Sand, soil, kieselguhr and silica gel

Sporulating fungi survive drying in soil (Fennell, 1960) and various species have been stored for periods of up to 5 years without change in characters (Atkinson, 1954). Selected yeasts, fungi and bacteria have been successfully dried on silica gel (Sidyakina, ch. 4.5, and Kirsop, ch. 6, this volume; Grivell and Jackson, 1969; Onions, 1971). Survival of bacteria for several years without change in characters has been reported by Grivell and Jackson (1969), but changes have been detected in some strains of yeast (Kirsop, this volume, ch. 6).

2. Paper strips or discs

Storage on paper has been successfully used with some yeasts (see Kirsop, this volume, ch. 6) and staphylococci (Coe and Clark, 1966). After drying the strips or discs may be stored in foil packets in airtight containers or between strips of self-adhesive plastic (Coe and Clark, 1966), providing an inexpensive method for the postage of large numbers of cultures.

3. Pre-dried plugs

Various materials such as starch, peptone or dextran have been used to make pre-dried plugs onto which small volumes of suspension of organisms are dropped before drying and storing under vacuum. This method has been successfully applied to delicate bacteria (*Neisseria gonorrhoeae* and *Vibrio cholerae*) which are difficult to freeze-dry (Annear, 1956, Malik, this volume ch. 4.10).

4. Gelatin discs

In this method, originally described by Stamp (1947), organisms are suspended in a nutrient gelatin medium and drops are allowed to solidify in petri dishes. The drops may be dried or freeze-dried and the resulting dried discs stored over silica gel or phosphorous pentoxide. Various additives have been used to supplement the nutrient gelatin. A variety of bacteria have been successfully preserved by this method with survival over several years. (Snell and Kocur, this volume, ch. 4.3; Obara *et al.*, 1981; Yamai *et al.*, 1979).

None of the above methods appear to be universally applicable and in most cases they have been used for particular groups of microorganisms or for rather specialist applications. It is therefore difficult to evaluate these methods fully in terms of the criteria discussed earlier, as in many cases the relevant information is not available. Long-term viability appears moderately good and can certainly be measured in years. Stability of strain characters has not always been examined and appears to be very strain-specific. Contamination is certainly likely to be less of a problem than with serial subculture. Capital equipment costs are small and none of the methods appears unduly labour-intensive in view of the expected survival times. Because of their technical simplicity, many of the methods would appear suitable for storing large numbers of cultures. Information on the reliability of these methods is restricted to a limited range of microorganisms and users would be well advised to gain first-hand experience before applying them to important cultures. Distribution would appear to be no problem as most of the methods lend themselves to batch production. Most of the methods appear specially suitable for storage of frequently used cultures since aliquots of dried material can be removed from containers without great risk of contamination.

C. Freeze-drying

Freeze-drying is a process in which water is removed by sublimation from the frozen sample. Organisms are suspended in a suitable medium, frozen and exposed to a vacuum. The water vapour removed is trapped either in a refrigerated condenser or phosphorous pentoxide. After drying the microorganisms are stored under vacuum or in an inert gas, most commonly in individual vials or ampoules. Two types of commercial freeze-dryer are in common use — the centrifugal and the shelf.

In the centrifugal dryer, the suspension is frozen by loss of latent heat associated with evaporation by vacuum. To increase the surface area and to avoid frothing of the suspension due to removal of dissolved gases before freezing is complete, the suspension is centrifuged during the initial stages of drying. Glass ampoules, which may be plugged, are used for centrifugal drying. After primary drying has been completed, the ampoule is constricted and placed upon a manifold for secondary drying. Vacuum is applied and further water is removed before flame-sealing the ampoule, either under vacuum or inert gas.

In the shelf dryer the suspension is pre-frozen before vacuum is applied. Freezing may be carried out by cooling the shelves in the freeze-dryer or by pre-freezing in a deep-freeze. Glass vials are used for shelf drying and may be stoppered automatically in the machine. It is not necessary to use a manifold for secondary drying as the complete drying programme is carried out continuously without further manipulation.

In general, centrifugal drying has achieved greater popularity than shelf drying. One advantage of centrifugal drying is that a cotton-wool plug can be inserted in the ampoule eliminating cross-contamination and acting as a filter to prevent scatter of microorganisms into the environment when the ampoules are opened. A further advantage is that the glass seal of ampoules is both air and moisture-tight for long periods, whereas vials are sealed with various formulations of rubber which may allow access of air and water vapour either through the natural permeability of the bungs or through leakage of the seals.

Freeze-drying has been widely used to preserve many different microorganisms (Lapage and Redway, 1974) and is widely applicable to yeasts, fungi, bacteria and some viruses. It is less applicable to algae and is unsuitable for protozoa, animal and plant cells.

A variation on freeze-drying termed 'L-drying' has been described (Annear, 1958; Lapage *et al.*, 1970; Dando and Bousfield, this volume, ch. 4.4). In this method, as in freeze-drying, suspensions of organisms are dried in ampoules under vacuum, but the vacuum is adjusted to allow rapid drying without freezing. The method has been used successfully with some species of bacteria that do not survive freeze-drying.

Although simple freeze-drying protocols can be established by trial and error, the process allows great refinement of control over a number of parameters and it is possible, if time permits, to optimize the process for the particular microorganisms to be preserved. The physical factors involved in the freeze-drying process have been discussed by Meryman (1966). The growth phase of the culture, temperature of growth, composition of suspending medium, rate of freeze-drying, final temperature of freezing, rate and duration of drying, and final moisture content are all factors which can be controlled and adjusted.

The advantages of freeze-drying are: suitability for batch production and distribution; maintenance of viability during storage (50 years or more for some microorganisms) without need for further attention; and undemanding storage requirements. For these reasons freeze-drying has found popularity in the service culture collections. The method suffers from some disadvantages. The capital cost is high if commercial equipment is purchased. Although a choice of machines is available commercially, equipment does not have to be sophisticated or unduly expensive since a vacuum pump, a manifold and a moisture trap are the only essential requirements. The freeze-drying process is fairly labour-intensive but, as large batches can be prepared, the labour hours per ampoule may be quite low.

Although, in general, stability of characters is good, some selection appears to take place during freeze-drying, and up to a thousand-fold drop in viability is not uncommon with more susceptible microorganisms. Selection may be amplified if serial batches are prepared, so that batch 2 is prepared from batch 1, batch 3 from batch 2, and so on. This effect can be avoided by reserving sufficient ampoules from batch 1 to act as a seed stock for subsequent batches. In addition to population changes, genetic change and loss of plasmids may also be experienced during freeze-drying of some species. From the user's point of view, freeze-dried cultures are time-consuming to open and resuscitate and several subcultures may be needed before organisms regain their usual morphological and physiological characteristics.

D. Freezing

In preservation by freezing, water is made unavailable to the microorganisms by freezing, and the dehydrated cells are stored at low temperatures. Damage may be caused to the cells both during the cooling stage and the subsequent thawing. This may be caused either by the concentration of electrolytes through the removal of water as ice or by the formation of ice crystals which may damage cellular integrity. Attempts to limit this damage may be made by adjustment of cooling and warming rates and the addition of various cryoprotectants such as dimethyl sulphoxide or glycerol to the cell suspension.

Methods can be broadly classed according to the storage temperature used. Temperatures of -20, -30, -40, -70, -140 and $-196°C$, have all been used but, in general, temperatures above $-30°C$ give poor results, due to the formation of eutectic mixtures exposing cells to high salt concentrations. Storage at $-70°C$ has been used for a variety of different microorganisms, including bacteria, fungi, mycoplasma, protozoa and viruses. Storage at $-140°C$ (nitrogen vapour phase) and $-196°C$ (nitrogen liquid phase) are being used increasingly, and successful results have been achieved for microorganisms and cultured cells that cannot be preserved in other ways.

1. Storage on glass beads at $-70°C$

A novel approach to storage at low temperatures was suggested by Feltham *et al.* (1978; see also Jones *et al.*, this volume, ch. 4.2). In this method, storage at $-70°C$ on small glass beads in a glycerol-suspending medium has been successfully used for a wide variety of bacteria. The method is very quick and easy to perform and requires no subsequent manipulation during storage. Each glass bead provides material for one subculture and allows large batches to be stored in minimal space. The method is ideally suited to storage of large in-house collections.

A disadvantage of the method is the high capital cost of a $-70°C$ freezer. In addition, provision must be made to safeguard against mechanical failure or prolonged interruption of electrical supply. Devices are available which provide automatic flushing with cold nitrogen vapour when a rise in temperature of the freezer occurs. Alternatively, a duplicate collection may be kept. The method is not well suited to frequent distribution of cultures as subcultures need to be made before issue.

2. Storage in liquid nitrogen

Storage in the liquid or vapour phase of nitrogen is the most universally applicable of all preservation methods. Fungi, bacteriophage, viruses, algae, protozoa, bacteria, yeasts, animal and plant cells and tissue cultures have all been successfully preserved. Although with some microorganisms, including genetically engineered strains, a high proportion of cells in a population may die on cooling and warming, and population changes may occur, virtually no further loss occurs on storage. Losses may be reduced by the use of cryoprotectants and adjustment of growth conditions and the rate of cooling and warming. For a detailed discussion of these factors see Morris (1981) and James (this volume, ch. 7.2). Although characters generally appear to be preserved without alteration, there have been some reports of nuclear and plasmid change (Calcott and Gargett, 1981; Williams and Calcott, 1982, Kirsop, this volume, ch. 6). Current knowledge suggests that longevity and stability of cultures with this method is higher for most cells than following freeze-drying, and storage in liquid nitrogen is nowadays the method of choice for preservation of valuable seed stock cultures. Storage in liquid nitrogen may be the only suitable method for long-term preservation of cells that will not survive freeze-drying.

The method has some disadvantages however. Liquid nitrogen evaporates and must be replenished regularly. Failure to do this, either through laboratory mishap or interruption of deliveries of liquid nitrogen through industrial action, may cause the loss of an entire collection. The capital cost of equipment is high, but the process is not labour-intensive. There is some risk of explosion if glass containers are kept in the liquid phase, as liquid nitrogen may penetrate through imperfect seals and expand rapidly when the container is warmed. Storage in the vapour phase removes this hazard, but the higher storage temperature may be considered less satisfactory for valuable strains. Alternatively, polypropylene containers may be used for cells other than strict anaerobes. The method is not very convenient for the distribution of cultures as subcultures have to be prepared. Storage space may become a problem particularly if working cultures are stored in addition to seed stocks, but various methods have been developed to reduce the storage space required (straws, Kirsop, this volume, ch. 6; glass beads, Jones *et al.*, this volume, ch. 4.2; glass capillaries, Hippe, this volume, ch. 4.8 and James, this volume, ch. 7.2).

For further general reading on maintenance methods see Hatt, 1980; Smith and Onions, 1983a and Sly, 1988.

BACTERIA

4.1. MAINTENANCE OF BACTERIA BY FREEZE-DRYING

R. H. RUDGE

National Collection of Type Cultures,
Central Public Health Laboratories, London, UK

I. Introduction 31
II. Details of method 32
   1. Pre-drying culture preparation 32
   2. Suspending fluids 32
   3. Preparation of suspension 32
   4. Preparation of ampoules 33
   5. Filling ampoules 33
   6. Primary drying 33
   7. Constricting ampoules 33
   8. Secondary drying 34
   9. Sealing 34
   10. Hazardous pathogens 34
   11. Vacuum testing of ampoules 35
   12. Opening ampoules 35
   13. Reconstitution 35
   14. Viability counts 35
III. Operation of the Modulyo freeze-dryer 36
IV. Storage conditions 37
V. General notes 37
VI. Organisms successfully preserved 38

I. INTRODUCTION

Freeze-drying, or lyophilization, is a process in which water vapour is removed directly from a frozen product by sublimation. It has been used for many years to preserve a wide variety of biological materials and among the numerous publications on the subject are reviews by Fry (1954, 1966), Heckly (1961, 1978), Hill (1981), Lapage and Redway (1974), Lapage et al. (1970) and Muggleton (1963), which cover freeze-drying of bacteria.

Unlike drying from the liquid phase, freeze-drying causes little shrinkage and results in a completely soluble product that is easily rehydrated. Chemical changes are minimized by preventing

MAINTENANCE OF MICROORGANISMS 2nd Edn
ISBN 012–410351–0

concentration of solutes and also by virtue of the lowered temperature, which reduces the rate of chemical reaction. One major advantage of freeze-drying over many other preservation methods is that material can be kept stable over a period of many years without the need for special storage conditions. Distribution of cultures is also simplified as no further preparation is necessary before despatch. There are many different methods leading to successful freeze-drying, but the techniques covered here are limited to those used by the UK National Collection of Type Cultures (NCTC).

II.   DETAILS OF METHOD

1. Pre-drying culture preparation

It is desirable that the best growth possible is obtained prior to drying and that it is in an easily harvested form. Thus cultures are grown in 150 × 19 mm tubes of a suitable non-selective, sloped medium, commonly nutrient or blood agar, using one tube per 10 ampoules to be dried. If necessary, liquid cultures can be used, though these need to be centrifuged in order to concentrate the cells before preparation of the drying suspension. The optimal stage of growth for harvesting may vary but, in general, late logarithmic phase cultures prove suitable.

2. Suspending fluids

(a) Inositol serum (Redway and Lapage, 1974)

meso-Inositol (Koch-Light)	5.0 g
Horse serum (Life Technologies Ltd)	100 ml

Sterilize by filtration and aseptically distribute 5 ml volumes into sterile bijoux bottles.

(b) Inositol broth

Oxoid Nutrient broth powder No. 2 (Unipath)	2.5 g
meso-Inositol (Koch-Light)	5.0 g
Distilled water	100 ml

Distribute 5 ml volumes into bijoux bottles and sterilize by autoclaving at 121°C for 15 min.

3. Preparation of suspension

A suitable suspending fluid is essential to prevent overdrying and to protect the bacteria from mechanical and chemical damage during both drying and storage. The NCTC uses 5% inositol serum routinely for all organisms except enterobacteria, for which 5% inositol broth is used in order to avoid possible immunological change or damage.

Harvesting is carried out by adding 1–2 ml of suspending fluid to each slope of culture and gently rubbing off the growth with a Pasteur pipette before emulsifying into a uniform suspension. This should be done carefully to avoid creating aerosols. With larger batches, the growth from several tubes is pooled prior to filling ampoules.

4. Preparation of ampoules
Neutral glass freeze-drying ampoules, 100 × 7.0–7.5 mm (Edwards High Vacuum), are acid-washed by soaking in 2% hydrochloric acid overnight, then thoroughly rinsed in tap water and finally distilled water. For each culture to be dried, labels are prepared by stamping or typing the required identification on blotting paper, and a 5 × 30 mm strip is placed in each ampoule. Ford's gold medal paper, 140 g/m² (Peter Bellingham Ltd), is used in the NCTC, and different colours are available for batch coding. Ampoules are then plugged with cotton wool and sterilized by autoclaving at 121°C for 15 min.

5. Filling ampoules
Ampoules are filled with the bacterial suspension using a Pasteur pipette; approximately 0.1–0.2 ml is delivered to the bottom of each ampoule, taking care not to contaminate the sides or top. As the ampoules are filled, the tops are flamed thoroughly and the cotton wool plugs replaced, though these are later discarded when all the ampoules are ready for loading on to the dryer.

6. Primary drying
This stage involves removal from the preparation of all the water that can be frozen. The suspensions are initially frozen by evaporative freezing under reduced pressure while centrifuging to prevent frothing due to evolution of dissolved gases. Because of the high rate of vapour flow during this process, obstructions such as cotton wool plugs should be avoided; instead, caps of gauze or cotton placed over each ampoule or group of ampoules, while not providing an absolute bacterial filter, will limit any contamination and contain any loose flakes of freeze-dried material. Primary drying is continued for a minimum of 3 h.

7. Constricting ampoules
On completion of primary drying, the ampoules are constricted to facilitate later sealing. It is advisable that this stage is carried out as quickly as possible to avoid prolonged exposure of the ampoules to air. Sterile plugs of non-absorbent cotton wool, previously autoclaved *in situ* in empty ampoules or tubes, are inserted approximately 15 mm into each ampoule and the top section, which has been handled, is cut off. The plug is pushed with a ramrod halfway down the ampoule and then, using a narrow flame and taking care not to char the cotton wool, the ampoule is constricted above the plug to produce a short capillary section of about 2 mm diameter. At the NCTC, this is done with

a 'Flair' handtorch [Jencons (Scientific) Ltd], using natural gas and oxygen to produce a very hot, slim flame. Alternatively, a fishtail burner or semi-automatic ampoule constrictor (Edwards High Vacuum) can be used.

8. Secondary drying

During this stage, the water that remains unfrozen, bound to the material by adsorption, is removed to leave a residual moisture content of around 1%. Compared with primary drying, desorption is a slow process and is often carried out using a chemical desiccant such as phosphorus pentoxide (P_2O_5) to trap the small amounts of water involved. Refrigerated vapour traps are less efficient but can be used provided a condenser temperature of below $-50°C$ is maintained.

9. Sealing

When secondary drying is complete, ampoules are sealed *in situ* on the dryer manifold while still under vacuum. The 'Flair' handtorch [Jencons (Scientific) Ltd] is well suited to this, but alternatively a crossfire burner (Edwards High Vacuum) or a small butane gas blowtorch may be used. If the constriction is too wide, or the flame held too close to the body of the ampoule, a hole may be produced in the ampoule as the melted glass is sucked inwards. Ampoules damaged in this way must be discarded.

It is also possible to seal ampoules at atmospheric pressure after first filling with a dry, inert gas, for example, nitrogen, introduced via the freeze-dryer air-admittance valve. However, subsequent integrity of such ampoules is not as easily checked as it is for vacuum-sealed ampoules.

10. Hazardous pathogens

Category 3 organisms, as listed in the UK by the Advisory Committee on Dangerous Pathogens (1990), together with some other hazardous pathogens, are dried at the NCTC by a slightly different method, which avoids evaporative freezing and centrifugation, and employs cotton wood plugs throughout the whole drying process. All operations are carried out in a containment level 3 laboratory, and ampoules are filled and subsequently opened in an exhaust protective cabinet.

After filling, ampoules are replugged with loose, non-absorbent cotton wool plugs, which are then trimmed and pushed halfway down the ampoule. Suspensions are quick-frozen by covering ampoules with crushed solid CO_2, and the frozen ampoules are transferred quickly to the freeze-dryer and subjected to a vacuum before they can thaw. The dryer incorporates a 0.2 μm 'Mini Capsule' pleated filter (Gelman Sciences Ltd) in-line between the condenser chamber and vacuum pump, to prevent any contamination of the pump or exhaust. Drying then proceeds as above, with the exception that the ampoules do not need to be plugged prior to constriction.

11. Vacuum testing of ampoules

For locating leaks in ampoules on the secondary drying manifold, and also for checking maintenance of vacuum in ampoules after storage, a useful item of equipment is a high-frequency spark tester (Edwards High Vacuum). This demonstrates a satisfactory vacuum by producing a pale-blue/violet glow within the ampoule, whereas a poor vacuum is indicated by a deep-purple glow or else no discharge at all. The tester should be used with care, however, avoiding the bottom and sealed tip of the ampoule, as these more fragile areas can be punctured by the spark.

12. Opening ampoules

Ampoules can be opened safely by making a score mark on the glass with a file or diamond at a point near the middle of the cotton wool plug and then applying a red-hot, fine glass rod or pipette tip to this mark. This should produce a crack encircling the ampoule; if only a short crack is produced, tap gently at this point to complete the encirclement. After allowing air to seep in, filtered by the plug, the tip of the ampoule can be removed; this and the plug are then discarded, treating both as infected material. The open end of the ampoule is flamed and a sterile cotton wool plug inserted.

13. Reconstitution

With a Pasteur pipette, approximately 0.5 ml nutrient broth is added to the ampoule and the contents mixed carefully to avoid frothing. With cultures known to give poor after-drying viability, it may be beneficial to leave the contents to rehydrate for a few minutes before transferring the suspension from the ampoule. The suspension is then used to inoculate suitable media including, if possible, an agar plate in order to detect any contaminants introduced during opening.

14. Viability counts

Each batch of cultures dried is checked routinely by counting the suspension before drying and again immediately after drying. This provides not only a viability check and a measure of the loss during drying, but also serves as a check for purity of the dried culture. Unless the count is low enough to warrant more frequent checks, further counts are made after 1 year, after 5 years and thereafter at 5-year intervals. Cultures are re-dried when the count falls below an acceptable limit.

Counts are performed using a modified Miles and Misra (1938) technique. The culture is rehydrated in 1 ml nutrient broth to provide a nominal 10^{-1} dilution, and from this a series of 10-fold dilutions is prepared, up to 10^{-6}. Using a pipette calibrated to deliver drops of 0.02 ml (Patterson Scientific), three drops of each dilution are delivered onto an agar plate of suitable medium. Results are generally expressed in terms of the highest dilution to yield growth, as this is adequate for the purpose of comparison, and there is no need to calculate the actual count per ampoule.

III. OPERATION OF THE MODULYO FREEZE-DRYER (EDWARDS HIGH VACUUM)

(1) Connect unit to mains power supply.

(2) Drain condenser by opening drain valve anticlockwise (valve and drain outlet at lower left of front panel). Close valve.

(3) Switch on condenser refrigeration unit (rocker switch by temperature gauge). Wait until condenser temperature falls below −40°C before commencing stage (7).

(4) Remove 'O' ring (where fitted) from condenser chamber flange.

(5) Position spin-freeze unit over condenser chamber. These units vary; however, with those incorporating a timer:
   (a) Place loaded ampoule carrier on centrifuge drive shaft.
   (b) Position chamber bell jar, with base L gasket in place, over ampoule carrier.
   (c) Lift 'spinner start' knob and turn to seat over chamber central ring.
   (d) Set timer to 10 min.
   (e) Connect unit to mains power supply, switch on (rocker switch at side of unit) and start centrifuge by pressing button in centre of timer.

(6) Close air admittance valve.

(7) Switch on vacuum pump. (NB. It is not essential to open the pump gas ballast valve during the drying of such small volumes — this is optional.)

(8) With those centrifuge units not incorporating a timer, switch off after approximately 10 min, by which time the vacuum should be below 2 mbar and the ampoules frozen.

(9) Continue drying for a minimum of 3 h, after which primary drying should be complete.

(10) Switch off vacuum pump and open the air admittance valve to admit air slowly.

(11) When the vacuum has been released, remove chamber bell jar and unload centrifuge head.

(12) Remove centrifuge unit from Modulyo and replace the 'O' ring (where fitted) on the condenser chamber flange.

(13) Switch off refrigerator and defrost condenser chamber by filling (not above level of pump line outlet) with warm water and leave until ice has melted before draining. Wipe condenser completely dry and close drain valve.

(14) Load the desiccant trays with phosphorous pentoxide, using a minimum of 5 g per millilitre of remaining water, and place tray assembly inside the condenser chamber.

(15) Position the acrylic cover plate over the condenser chamber flange (secure with three screws on earlier models), with the secondary drying manifold(s) fitted to the centre coupling on this plate.
(16) Plug and constrict ampoules.
(17) Place constricted ampoules on manifold, twisting the ampoules slightly to ensure a good seal. If necessary, use empty ampoules to completely fill the manifold.
(18) Close air admittance valve and switch on vacuum pump.
(19) Ensure that a Pirani gauge reading of $<10^{-1}$ mbar is attained; if not, check ampoules for leaks using a high frequency spark tester and if necessary replace faulty ampoules with blanks.
(20) Continue secondary drying overnight (15–20 h), then flame-seal ampoules under vacuum.
(21) Switch off vacuum pump and open air admittance valve.

If it is preferred to carry out secondary drying using the refrigerated condenser as a vapour trap in place of phosphorous pentoxide, omit stages (13) and (14).

IV.   STORAGE CONDITIONS

Stability of freeze-dried cultures stored in the dark at normal room temperatures is generally very good, although there may be some decrease in shelf life if ambient temperatures are consistently high. This difference can be utilized in the accelerated storage test, in which the viability loss of cultures held at high temperatures is used to predict the shelf life at normal storage temperature (Mitic *et al.*, 1974). For most cultures, storage at room temperature will prove adequate, though some of the more delicate organisms (see Section VI) may benefit from storage in a refrigerator or freezer.

V.   GENERAL NOTES

(1) Used P_2O_5 is best disposed of by first leaving to hydrate completely by exposure to air. The sludge that remains can then be dissolved by soaking in hot water before diluting and running to sink waste.
(2) Pyrex or borosilicate glass ampoules should not be used for freeze-drying. Although they are stronger, they can be extremely difficult to open.

(3) When preparing ampoules, and again when filling, the tops should be examined for cracks or irregularities, and any flawed ampoules discarded as these can splinter or leak when placed on the secondary drying manifold.

(4) The rubber manifold nipples can be *lightly* greased periodically, using Apiezon high vacuum grease (Edwards High Vacuum) to facilitate a good seal with the ampoules.

(5) Ampoule identification labels should include a 10 mm gap to the left of the number, and this end is placed towards the bottom of the ampoule. This prevents the dried material obscuring the number and helps to ensure that the figures are always read from the same end and thus the right way up.

VI. ORGANISMS SUCCESSFULLY PRESERVED

Table I, compiled from records of viability counts over the past 35 years, lists bacterial genera that are routinely dried at the NCTC. Figures given for the mean logarithmic count represent an average value of results obtained for a given number of strains in each genus and indicate the highest dilution to give single figure colony counts. Not all of the 4000+ NCTC strains are included in these figures, only those from each genus that have been counted over the longest period. Although some batches that were discarded early because of low counts are therefore omitted, this bias is somewhat compensated for by the inclusion of older batches that were dried and stored under sub-optimal conditions.

The majority of bacteria survive freeze-drying well, but a few species can sometimes give disappointing results. This may be due in some cases to difficulties in obtaining adequate predrying growth. Cultures which often prove more difficult than others include *Aquaspirillum serpens, Clostridium botulinum, C. chauvoei, C. novyi, C. putrificum, C. scatologenes, Helicobacter pylori* and *Peptococcus heliotrinreducans.* Additionally, some lesser problems may be encountered with *Bacteroides melaninogenicus, Haemophilus canis, H. suis, Leptotrichia buccalis, Mycobacterium microti* and *Neisseria gonorrhoeae.*

Table I. Survival (mean logarithmic counts) of bacteria freeze-dried at the UK National Collection of Type Cultures before (BD) and after drying (AD) and after storage for various periods.

Genus	No. of species	No. of strains	BD	AD	Storage period (years)							
					1	5	10	15	20	25	30	35
Achromobacter	6	9	7.0	7.0	6.9	6.8	6.0	(5.5)	+			
Acinetobacter	2	10	7.0	7.0	6.9	6.8	6.7	(6.6)	(6.0)	+		
Actinobacillus	3	19	6.3	6.1	5.6	4.9	4.2	3.7	(3.5)	+	+	
Actinomadura	3	8	4.6	4.5	4.4	4.3	4.0	+	+			
Actinomyces	4	13	5.2	4.9	4.5	4.2	3.8	(3.2)	(3.2)	(2.7)	+	+
Aerococcus	1	14	5.8	5.7	5.6	5.6	5.5	5.5	5.2	(5.3)	+	
Aeromonas	4	9	7.0	6.7	6.7	6.2	6.2	+	+			
Alcaligenes	3	7	7.0	6.9	6.6.	6.6	(6.4)	.(6.0)	+			
Alteromonas	3	8	6.9	6.8	6.6	6.0	(5.5)	(5.3)				
Alysiella	1	1	5.0	5.0	5.0	5.0	5.0	4.0				
Anaerobic coccus	–	11	6.2	5.7	5.2	4.8	4.5	3.8	(3.6)	+	+	
Anaerobiospirillum	1	1	5.0	4.0	4.0	4.0	2.0					
Aquaspirillum	1	1	7.0	4.0	3.0	2.0	2.0	2.0				
Bacillus	27	70	5.8	5.5	5.4	5.2	5.0	4.8	4.5	(4.7)	(4.1)	+
Bacterionema	1	3	5.3	5.0	5.0	4.0	3.0					
Bacteroides*	5	12	6.4	6.0	5.2	4.4	3.6	(3.2)	+			
Beneckea	4	4	6.3	5.8	4.8	3.5						
Bifidobacterium	1	2	6.5	6.5	6.5	6.0	6.0					
Bordetella	3	26	7.0	6.9	6.7	6.6	6.4	(6.3)	(6.2)	+		
Brachyspira	1	1	5.0	4.0	3.0							
Branhamella	2	8	6.0	5.8	5.8	5.5	5.5					

Table I. Continued.

Genus	No. of species	No. of strains	BD	AD	Storage period (years)							
					1	5	10	15	20	25	30	35
Brevibacterium	2	4	7.0	7.0	7.0	7.0	(7.0)					
Brochothrix	1	1	6.0	6.0	6.0	6.0						
Brucella	5	32	6.8	6.8	6.8	6.5	6.2	5.8	(5.7)	(5.7)	+	
Buttiauxella	1	1	6.0	6.0	6.0							
Butyribacterium	1	1	6.0	6.0	6.0	6.0	6.0	6.0				
Campylobacter	13	43	6.3	5.4	5.1	(5.0)	(4.7)	+				
Capnocytophaga	3	7	6.1	5.9	5.4	+						
Cardiobacterium	1	3	7.0	6.3	5.7	5.3	5.0					
Cedecea	3	3	7.0	7.0	6.7							
Cellulomonas	2	2	6.5	6.5	6.5	6.5	6.0	5.0	5.0			
Chromobacterium	3	19	6.7	6.1	5.5	5.2	4.8	4.2	(4.0)	+		
Chryseomonas	1	3	6.3	5.7	5.3							
Citrobacter	2	11	7.0	6.9	6.8	6.6	6.5	6.0	(6.1)	(6.0)	+	
Clostridium*	18	73	5.0	4.6	4.5	4.2	3.9	3.8	3.6	(3.3)	+	
Comamonas	1	3	7.0	6.7	6.7	6.3	5.7	(6.0)	+			
Corynebacterium	14	72	6.4	6.1	5.9	5.8	5.4	5.1	4.9	(4.9)	(4.3)	+
Cytophaga	1	1	7.0	6.0	6.0	6.0	6.0					
Dermatophilus	1	1	4.0	4.0	4.0	4.0						
Edwardsiella	3	5	6.4	6.2	6.0	6.0	6.0	(6.0)	(5.0)			
Eikenella	1	2	6.0	6.0	5.5	4.5	(4.0)	(3.0)	(2.0)			
Enterobacter	8	20	6.8	6.7	6.6	6.5	6.3	6.1	(5.9)	(5.8)	(5.7)	
Erwinia	1	4	7.0	7.0	7.0	7.0	7.0	6.5	(6.0)			

Table I. *Continued.*

Genus	No. of species	No. of species	BD	AD	Storage period (years)							
					1	5	10	15	20	25	30	35
Erysipelothrix	1	8	6.3	6.1	6.0	5.9	5.6	5.4	5.2	+		
Escherichia	3	36	6.9	6.5	6.3	5.9	5.8	5.6	5.3	5.0	(4.4)	+
Eubacterium	3	3	6.3	6.3	6.3	6.3						
Ewingella	1	1	6.0	6.0	6.0							
Flavobacterium	11	26	6.9	6.7	6.6	6.6	(6.5)	(6.2)				
Francisella	1	1	6.0	6.0	6.0		5.0					
Fusobacterium	6	10	6.7	6.5	6.2	6.0	(5.4)	+				
Gardnerella	1	3	6.0	5.7	5.3	5.0	(5.0)					
Gemella	1	3	6.0	6.0	6.0	6.0	5.0	4.5				
*Haemophilus**	12	25	6.3	5.9	5.3	4.9	4.1	(3.3)	+	4.6	(4.2)	+
Hafnia	1	8	7.0	6.4	6.0	5.6	5.3	4.9	4.6	4.6		
Kingella	3	8	5.3	5.3	5.0	5.0	4.2	(3.7)				
Klebsiella	6	57	6.8	6.8	6.7	6.6	6.5	6.1	6.0	(5.8)	+	
Kluyvera	2	2	7.0	7.0	7.0	7.0	7.0					
Koserella	1	1	7.0	7.0	7.0							
Kurthia	2	3	6.3	5.7	5.3	5.0	4.3					
Lactobacillus	3	6	5.7	5.3	5.2	5.0	4.7	(4.6)	(4.6)	+		
Legionella	10	19	6.3	5.8	5.6	5.2	+					
Leminorella	2	2	7.0	7.0	7.0	7.0						
Leuconostoc	1	1	6.0	6.0	6.0	6.0	6.0	6.0				
Levinea	1	2	7.0	7.0	7.0	7.0	6.5	6.0				

Table I. Continued.

Genus	No. of species	No. of strains	BD	AD	Storage period (years)							
					1	5	10	15	20	25	30	35
Listeria	6	19	6.3	6.3	6.1	6.0	5.8	5.6	(5.5)			
Micrococcus	3	23	6.1	6.0	6.0	6.0	6.0	6.0	5.7	5.7	(5.4)	(5.3)
Mitsuokella	1	1	6.0	4.0	4.0							
Mobiluncus	2	9	6.3	6.0	5.8	5.5						
Moellerella	1	1	7.0	7.0	7.0							
Moraxella	12	25	6.0	6.0	5.8	5.4	5.4	(5.2)	+	+		
Morococcus	1	1	5.0	4.0	4.0	4.0						
Mycobacterium*	12	49	5.6	5.4	5.2	5.0	4.6	4.5	4.3	(4.1)	(3.5)	+
Mycococcus	2	.2	6.5	6.5	6.5	6.5	6.5	6.5	6.0			
Neisseria*	9	41	6.6	6.1	5.6	4.9	4.1	3.1	(2.8)	(2.4)	(2.1)	
Nocardia	5	10	6.1	6.0	5.9	5.9	5.7	+	+	+		
Ochrobactrum	1	5	6.6	6.6	6.6	6.6						
Oerskovia	1	1	7.0	7.0	7.0							
Oligella	2	3	6.3	6.3	6.3							
Pasteurella	5	28	6.9	6.8	6.2	5.8	5.6	5.4	5.2	(4.4)	+	
Pediococcus	1	2	6.0	6.0	6.0	6.0	6.0					
Peptococcus*	7	8	6.0	5.9	5.5	5.5	+					
Peptostreptococcus	3	3	6.0	5.7	5.0	4.7						
Plesiomonas	1	3	7.0	6.0	6.0	5.7	5.7	5.0	4.4			
Propionibacterium	4	6	6.7	6.7	6.5	6.5	6.3	6.5	+	+		
Proteus	4	22	6.7	6.6	6.6	6.6	6.5	6.5	(6.5)	(6.3)	+	
Providencia	3	11	7.0	6.8	6.7	6.4	6.3	6.0	(5.6)	+	+	

Table I. *Continued.*

Genus	No. of species	No. of strains	BD	AD	Storage period (years)							
					1	5	10	15	20	25	30	35
Pseudomonas	12	31	6.8	6.4	6.1	5.5	5.0	4.5	(4.3)	(3.7)	+	
Ramibacterium	1	1	6.0	6.0	6.0	5.0	5.0					
Rhodococcus	5	14	6.5	6.5	6.4	6.3	6.1	6.0	(5.6)	(5.0)	(4.6)	(4.3)
Rothia	2	3	6.0	6.0	6.0	6.0	6.0	6.0				
Salmonella	—	131	7.0	6.5	6.2	5.8	5.4	5.1	4.9	4.7	4.4	4.3
Sarcina	1	1	7.0	7.0	7.0	7.0	7.0	7.0				
Serratia	6	18	6.9	6.9	6.9	6.8	6.7	6.5	(6.2)	(6.0)	+	
Shigella	4	63	6.9	6.7	6.5	6.1	5.8	5.6	5.5	5.2	5.0	(4.6)
Simonsiella	1	1	4.0	4.0	3.0	3.0	3.0					
Staphylococcus	3	48	6.3	6.1	6.1	6.0	5.9	5.8	5.6	5.3	(5.2)	(4.7)
Streptobacillus	1	8	4.5	4.3	4.0	3.5	+					
Streptococcus	12	75	6.0	5.8	5.7	5.5	5.1	5.0	4.7	4.5	4.1	(3.6)
Streptomyces	5	9	4.1	3.7	3.7	3.6	3.5	3.3	(3.0)	+		
Thermoactinomyces	1	1	4.0	4.0	4.0	4.0	4.0	4.0				
Veillonella	8	9	6.8	6.3	6.1	6.0						
Vibrio	5	23	7.0	6.6	6.0	5.4	5.0	4.6	(4.0)	+	+	
Weeksella	2	8	6.1	6.0	5.9	5.6	5.5	+				
Wolinella	3	3	6.7	6.0	6.0	6.0						
Yersinia	2	20	6.6	6.6	6.4	6.4	6.2	5.9	(5.3)	(4.9)		
Zooglea	1	1	7.0	7.0	7.0	6.0	5.0					

Count figures in parentheses are based on fewer than the indicated number of strains tested.

+ Indicates cultures still viable, though an insufficient number of strains tested to provide a representative log count.

* Some species in these genera may prove difficult to freeze-dry and result in a relatively poor survival rate.

4.2. MAINTENANCE OF BACTERIA ON GLASS BEADS AT −60°C TO −76°C

D. JONES, P. A. PELL and P. H. A. SNEATH

Department of Microbiology,
University of Leicester, Leicester, UK

I.	Introduction	45
II.	Method	46
	1. Batch size	46
	2. Preparation of beads	46
	3. Preparation of sterilized vials	47
	4. Preparation of suspending medium	47
	5. Growth of bacteria	47
	6. Labelling of vials	48
	7. Preparation of bacterial suspension	48
	8. Distribution of suspension	48
	9. Freezing and storage of material	48
	10. Recovery and checking of frozen bacteria	49
III.	Storage conditions	49
IV.	Organisms successfully preserved	50
V.	Shelf life	50

I. INTRODUCTION

Storage of cultures in the range of −60°C to −80°C is possible in many laboratories because of the ready availability of commercial deep-freezers within this temperature range. The disadvantage of storing bacteria in this way is the damage caused by repeated freezing and thawing when subcultures are required. To overcome this problem a method based on the use of frozen bacterial suspensions with a cryoprotectant in glass beads was developed in our laboratory (Feltham *et al.*, 1978). The technique allows individual beads to be removed without thawing the whole sample. After more than 10 years of use in our laboratory and in other laboratories, the method has proved to be a safe, reliable and simple procedure for the storage of a wide range of bacteria

MAINTENANCE OF MICROORGANISMS 2nd Edn
ISBN 012–410351–0

(Feltham *et al.*, 1978; Pell and Sneath, 1984). The advantages of the method are:

(1) minimum preparation of materials is required;
(2) the method is simple to perform;
(3) many hundreds of strains can be stored for long periods in a small space;
(4) recovery of cultures is quick with little or no disturbance to other stored cultures;
(5) only the portion of culture removed is thawed; the bulk of the stock culture remains frozen;
(6) the beads thaw rapidly when placed on solid growth medium and recovery is immediate;
(7) with a suspending medium of suitable composition, the method can be used for most aerobic and anaerobic bacteria;
(8) stability of phenotype appears comparable to that achieved by freeze-drying;
(9) beads of different colours can be used to identify various categories of bacteria;
(10) viability and stability of cultures is not seriously impaired by breakdown of refrigeration for a few days.

II. METHOD

The steps involved in preparing the bacterial suspensions are:

(1) sterilization of vials containing washed beads;
(2) growth of bacteria;
(3) suspension in an appropriate medium containing a cryo-protectant, such as glycerol;
(4) distribution of suspension to beads in vials;
(5) freezing of vials;
(6) checking of product.

1. Batch size   The batch size prepared will depend on the intended use of the culture. About 20–30 prepared glass beads are placed in vials of approximately 2 ml capacity. One culture may be distributed in more than one vial, and it is advisable to prepare at least two vials per culture — one may then be used for routine recovery, the other as a reserve in case the first becomes contaminated or yields no growth. If two freezers are available, one vial can be put in each, to guard against loss.

2. Preparation   Glass, 2 mm embroidery beads (Creative Beadcraft Ltd) are
of beads   washed in tap water with a detergent, followed by dilute HCl to

neutralize alkalinity. The beads are then washed several times in tap water until the pH of the wash water is that of tap water. The beads are finally washed in distilled water, then dried at 45°C in an oven.

The suppliers produce beads of various colours, which may be used to differentiate various groups of bacteria according to the requirements of a particular laboratory; for example: animal pathogens, red; plant pathogens, green; special growth conditions required, blue; teaching strains, white.

3. Preparation of sterilized vials

About 20–30 prepared beads are placed in screw-cap glass vials of 1.75 ml capacity (Philip Harris Scientific). The vials are capped and sterilized by autoclaving at 121°C for 15 min. Quantities of autoclaved vials may be stored until required.

4. Preparation of suspending medium

For aerobic bacteria, 10 ml quantities of 15% (v/v) glycerol in nutrient broth (Difco Laboratories) are prepared in Universal bottles and sterilized by autoclaving at 121°C for 15 min. For anaerobic bacteria, BGP medium (Barnes, 1969) without agar but with 15% (v/v) glycerol is recommended. This medium is of the following composition:

Tryptone	(Unipath Ltd)	10.0 g
NaCl		5.0 g
Beef extract	(Lab-Lemco, Unipath)	3.0 g
Yeast extract	(Difco)	5.0 g
Cysteine hydrochloride		0.4 g
Glucose		1.0 g
Na_2HPO_4		4.0 g
Glycerol		150 ml
Distilled water		1000 ml

Dispensed in 10 ml quantities in Universal bottles.
Sterilized by autoclaving at 121°C for 15 min.

To freeze halophiles or alkalophiles the appropriate growth medium plus 15% (v/v) glycerol has proved satisfactory.

5. Growth of bacteria

Bacteria should be grown on the most appropriate, non-selective, solid medium under the optimum growth conditions. The use of a solid, non-selective medium reduces the risk of contamination. The number of growth plates used for any culture depends on the batch size required and on the vigour of the organism. Experience has shown that loss of viability after repeated sampling of one vial is reduced markedly when thick bacterial suspensions ($>10^8$ organisms ml^{-1}) are used.

6. Labelling of vials   Sterile vials are labelled for each organism. The appropriate culture collection number, the most suitable medium for recovery, date of preparation of material, and any other pertinent information is written on a self-adhesive label and stuck on the side of each vial. It is recommended that these labels be further secured by wrapping a layer of clear sticky tape completely around each vial. This method of labelling is less expensive than the special deep-freezer tapes. The black caps may be whitened with waterproof ink and the culture collection number written on the top with waterproof ink. This facilitates detection and retrieval of the appropriate vial when the organism is required.

7. Preparation of bacterial suspension   Strains incubated overnight under suitable conditions should be carefully inspected for contaminants. Approximately 1 ml of the appropriate sterile suspending medium is aseptically pipetted onto the plate and, using a wire loop, the growth is emulsified with the broth to make a thick suspension.

8. Distribution of suspension   With a sterile Pasteur pipette the bacterial suspension is aseptically dispensed into each of the two prepared vials. The suspension should be aspirated several times to ensure the air bubbles inside the beads are displaced by the bacterial suspension. After the beads are thoroughly wetted, the excess suspension should be removed from the bottom of the vial. Excess suspension left in the vial makes it more difficult to remove individual beads when required after storage.

An alternative method for mixing bacteria with the beads is to wet the beads with the appropriate suspending medium, then agitate one loopful of the growth from the plate amongst the beads. Again, this should be done so that the air bubbles are displaced. This method is especially useful when only slight bacterial growth is available.

9. Freezing and storage of material   The vials are placed in trays of suitable size. We have used anodized aluminium sectioned trays (Denley Instruments Ltd). However, smaller trays could prove more convenient, and a tray with a lid prevents accumulated freezer ice falling on the vials during manipulation.

The trays are placed in a commercial freezer capable of maintaining temperatures of $-60°C$ to $-80°C$. A temperature of $c. -70°C$ is recommended. Removal of single beads after storage is facilitated if the vials are frozen on their side or if the beads are tapped onto the side of the vial before freezing in the upright position.

If a number of bacterial cultures are stored by this method it is recommended that a record of the contents of each tray and the position of each tray in the freezer is recorded in a culture collection notebook. This makes location and retrieval of cultures easier.

10. Recovery and checking of frozen bacteria

As with any method of preservation the bacteria should be checked for viability, purity and retention of particular characteristics after freezing.

The vial is removed from the freezer and one bead removed using a mini-spatula sterilized by flaming in alcohol and then cooling. The vial should be replaced immediately to prevent the remaining contents from thawing. The bead is rubbed over the surface of a suitable solid medium with a wire loop so that the bacterial inoculum is released. The plate is then incubated under appropriate conditions.

If a number of vials are removed at one time or if the freezer is located at some distance from the work-bench, thawing may occur during transit. This problem can be minimized by immersing the vials in boxes containing solid CO_2 or by using a cold block of paraffin wax containing holes into which the vials fit. We favour the block of paraffin wax. This is made by pouring paraffin wax into a tin box; vertical holes of a size suitable to contain the vials are then drilled part way through the block. The block is kept in the freezing cabinet. Vials for sampling may be placed in the cavities and the block of wax carried to the laboratory bench; the cold wax keeps the vials frozen for up to an hour. Long-term storage of cultures in solid CO_2 causes a lowering of culture pH, and this could be detrimental.

III. STORAGE CONDITIONS

The vials are stored in a commercial freezer at temperatures between $-60°C$ and $-76°C$. Within this range, the actual temperature does not appear to be very important.

With any refrigeration unit there is always the risk of breakdown and emergency back-up facilities are expensive. However, repairs are usually conducted in a matter of hours or, at worst, a few days. Recent work (Pell and Sneath, 1984) indicates that bacteria frozen on beads can survive such breakdowns, and those tested remained viable for a few days. It is recommended that if thawing has been in progress for more than a few hours, the collection should be re-preserved with newly grown bacteria.

IV. ORGANISMS SUCCESSFULLY PRESERVED

We have successfully preserved various species of *Actinobacillus*, *Clostridium*, *Haemophilus*, *Neisseria*, *Pasteurella*, *Yersinia*, *Vibrio*, a number of Enterobacteriaceae, staphylococci, micrococci, lactobacilli, streptococci and coryneform bacteria. Other laboratories have successfully preserved alkalophiles, halophilic archaebacteria, propionibacteria and *Bacteroides* spp. The method has been in use for some 12 years, but not all bacteria have been preserved for this length of time.

V. SHELF LIFE

In our experience to date, good levels of viability are maintained for over 10 years. There is every reason to believe that the 'shelf life' is, in fact, a good deal longer. Compare also with Chapter 4.4, V.

4.3. MAINTENANCE OF BACTERIA IN GELATIN DISCS

J. J. S. SNELL and M. KOCUR[1]

Division of Microbiological Reagents and Quality Control,
Central Public Health Laboratory, London, UK
[1] Czechoslovak Collection of Microorganisms,
Brno, Czechoslovakia

I.	Introduction	51
II.	Details of method	52
	1. Batch size	52
	2. Growth of bacteria	52
	3. Harvesting	53
	4. Gelatin suspending medium	53
	5. Distribution of drops and freezing	53
	6. Freeze-drying	53
	7. Preparation of vials	53
	8. Distribution of discs to vials	54
	9. Revival of dried bacteria	54
	10. Checking the product	54
III.	Storage conditions	54
IV.	General notes	55
V.	Successfully preserved organisms: shelf-life	55

I. INTRODUCTION

Preservation of bacteria in the form of gelatin discs was first described by Stamp (1947). A harvest of bacterial growth is suspended in melted nutrient gelatin, drops of which are allowed to solidify in Petri dishes. The drops are dried, or freeze-dried, over a desiccant and the resultant flat discs are stored over silica gel. For use, a single disc is placed in warmed broth and the resulting suspension plated onto a suitable growth medium. The method is not particularly suitable for storage of numerous strains over long periods; however, it is invaluable for storage of a limited number of frequently used strains, such as those used for quality control of media or regeants. The essential advantages of the method are:

MAINTENANCE OF MICROORGANISMS 2nd Edn
ISBN 012–410351–0

(1) ease of use;

(2) ease of storage — 30 or 40 discs can be kept in a 14 mm screw-capped vial;

(3) freedom from contamination — as the discs are kept dry there is no opportunity for growth of any contaminants introduced during sampling;

(4) stability of characters — as the bacteria are not growing there is no opportunity for mutation and selection.

The method therefore has advantages over both active subculture on slopes and freeze-drying in ampoules.

A number of organizations (American Type Culture Collection, Difco Laboratories, Remel) provide standard strains of micro-organisms in this form at a cheaper rate than freeze-dried cultures. In addition, the Czechoslovak Culture Collection (CCM) makes available 20 different strains for control purposes or for use in diagnostic laboratories. It also prepares discs as a service to customers.

II. DETAILS OF METHOD

The steps involved in preparing the discs are:

(1) growth of bacteria;

(2) suspension in gelatin;

(3) distribution of drops to Petri dishes;

(4) freezing the drops;

(5) freeze-drying the drops;

(6) distribution of dried discs to vials;

(7) checking the product.

1. Batch size   The batch size prepared will depend on the intended use and distribution of the discs. As a guide, the base of one 9 cm Petri dish will accommodate about 80 discs. This number should be more than enough for a year's supply for the average user. As a guide to scaling up the operation, an Edwards EFO3 freeze-dryer will accommodate a batch size of about 5000 discs dried in Petri dishes. All volumes given in this method are for the single Petri dish load of about 80 discs.

2. Growth of bacteria   Bacteria may be grown on any suitable non-selective media. Nutrient or blood agar will be suitable for many strains. A single 150 × 19 mm tube of sloped medium will provide adequate growth of bacteria such as Enterobacteriaceae or staphylococci.

3. Harvesting   Growth is harvested with a Pasteur pipette in a minimal volume (about 0.5 ml) of nutrient broth and added to 3 ml of the gelatin suspending medium previously melted and held at 37°C. This is mixed well to suspend.

4. Gelatin-
suspending
medium

Gelatin powder	(Unipath Ltd)	10.0 g
Nutrient broth powder, No. 2	(Unipath Ltd)	2.5 g
meso-Inositol	(Koch-Light Laboratories)	5.0 g
Deionized water		100 ml

Dissolve the solids by gentle heating, check the pH and adjust if necessary to 7.2. Distribute to screw-capped 6 ml bottles (bijoux) in 3 ml volumes and sterilize by autoclaving at 121°C for 15 min.

The CCM has found that the addition of skimmed milk to the basal medium improves both survival and the quality of the discs.

5. Distribution
of drops and
freezing

With a dropping pipette delivering 0.02 ml ('fifty dropper'), drops of the suspension are placed in the base of a plastic Petri dish. Maximum use of the area can be obtained by adopting a spiral pattern, starting from the outside. With care, about 80 drops can be accommodated in the base. The base of the Petri dish is covered with the lid and carefully placed in a deep-freeze at −20 to −40°C until the drops are frozen. Freezing is indicated by a change in appearance from transparent to opaque. It will occur in about 20 min with a light load, but may take up to 2 h with a large batch. Drops freeze fastest in the Petri dish at the bottom of the stack and dishes should be periodically rotated throughout the stack.

6. Freeze-
drying

The Petri dishes are transferred quickly to the freeze-dryer. If necessary, three piles can be accommodated under the plastic dome of the Edwards EFO3. The desiccant trays of the dryer must previously have been loaded with phosphorus pentoxide. The freeze-dryer is switched on and the cultures are dried overnight. With large numbers of discs in a batch it may be necessary to replace the phosphorus pentoxide after 2–4 h. This may be done by isolating the drying chamber, switching off the machine and venting the trap before replacing the P_2O_5.

7. Preparation
of vials

Coarse, self-indicating silica gel (BDH Chemicals Ltd) is placed in 14 × 45 mm screw-necked vials (FBG Trident Ltd) to a depth of about 10 mm and packed down tightly with a wad of cotton wool. The vials are sterilized in a covered container in a hot-air oven at 160°C for 1 h. The caps are sterilized separately by autoclaving at 121°C for 30 min. Caps are dried in an oven at 60–80°C for 4 h before placing on bottles.

8. Distribution of discs to vials

With small batches, discs may be transferred to vials with a small spatula. During drying the discs become detached from the plastic surface of the Petri dish and are easily picked up with a spatula. When a large batch is prepared discs can be transferred to vials with a small, wide-necked funnel. Normal clean technique should be observed in distribution; strict asepsis is not necessary since any airborne contaminants will have no opportunity to multiply on the dried discs. After distribution, the caps of the vials should be replaced and tightened.

9. Revival of dried bacteria

With fine-nosed forceps one gelatin disc is placed in 1 ml of nutrient broth. The broth is warmed in a 37°C incubator until the disc dissolves. A loopful of the broth suspension is transferred to a suitable solid medium and streaked to obtain single colonies before incubating.

10. Checking the product

Bacteria preserved by this method must be checked for viability and for retention of the particular characteristics for which they have been preserved. As this method is unlikely to be used as a sole means of preserving important cultures, it is probably not necessary to perform viable counts on the discs as simple plating will give a good indication of the level of viability. As in any method of preservation, it is essential to characterize the strain after drying to ensure that the correct strain has been preserved and has retained its important characteristics. After 4 years' experience in the use of this method, the CCM has found little change in phenotypic characters of strains used for quality control or in identification kits.

III. STORAGE CONDITIONS

Discs are stored at 5°C. It is important that the vials are allowed to warm to room temperature before opening to prevent condensation of water in the vial.

The CCM has found that storage at room temperature in the dark did not greatly affect the survival of cultures after 2 years. Obara *et al.* (1981) stored gelatin discs at −20°C.

IV. GENERAL NOTES

(1) Vented Petri dishes should be used to allow escape of water vapour during drying.

(2) In warm weather the gelatin drops may not set after dropping and will coalesce when moved. This may be overcome by placing the Petri dish on a layer of ice during the dispensing of the drops.

(3) If large batches are to be prepared drops can be placed in the base and the inside of the lid of the Petri dish. The drops in the lid are frozen before replacing the lid on the Petri dish.

(4) Distribution of the discs to vials is greatly facilitated if the discs have become detached from the surface of the Petri dish. For unknown reasons, discs in Petri dishes placed on the base plate of the freeze-dryer sometimes remain attached to the Petri dish. A dummy layer of empty Petri dishes on the base plate solves this problem.

V. SUCCESSFULLY PRESERVED ORGANISMS: SHELF-LIFE

Various species of Enterobacteriaceae and Staphylococci, and strains of *Pseudomonas aeruginosa* and *Corynebacterium diphtheriae* have been successfully preserved for at least 4 years. The method has not been successful with more delicate species such as *Neisseria* or *Haemophilus*. However, Obara *et al.* (1981) using a method described by Yamai *et al.* (1979) have reported successful preservation of *Neisseria*, *Haemophilus* and *Bacteroides*, using a gelatin-disc method based on a different suspending mixture.

BACTERIA

4.4. MAINTENANCE OF INDUSTRIAL AND MARINE BACTERIA AND BACTERIOPHAGES

T. R. DANDO and I. J. BOUSFIELD

NCIMB Ltd,
Aberdeen, UK

I. Introduction 57
II. Freeze-drying 58
   A. Suspending medium.......................... 58
   B. Preparation of 'mist. desiccans' 58
III. L-Drying 59
   A. Method 59
   B. Shelf-life 60
IV. Microdrying 61
   A. Method 61
   B. Shelf-life 61
V. Freezing over liquid nitrogen 62
   A. Bacteria..................................... 62
      1. Method.................................. 62
      2. Shelf-life 62
   B. Bacteriophages 63
      1. Method.................................. 63
      2. Shelf-life 63

I. INTRODUCTION

There are a number of different methods for maintaining bacteria. The methods described in this chapter have been successfully used in the National Collections of Industrial and Marine Bacteria (NCIMB) for the maintenance of a wide range of species.

MAINTENANCE OF MICROORGANISMS 2nd Edn
ISBN 012–410351–0

II. FREEZE-DRYING

The principal method in the NCIMB for the maintenance of the majority of bacterial species is a standard freeze-drying (lyophilization) procedure essentially the same as that described by Rudge (this volume, ch. 4.1). There are variations in procedures, however. For instance, in the NCIMB the ampoules are prepared containing the filter paper slip as described by Rudge, but lint caps are used in place of cotton-wool plugs. The latter are inserted just prior to the ampoules being constricted. In addition, the ampoules are placed in aluminium racks, and wrapped in greaseproof paper before sterilization.

A. Suspending medium

An important feature of the NCIMB method is the use of '*mist. desiccans*' (Fry and Greaves, 1951) as the suspending medium. This is a mixture of horse serum (Life Technologies Ltd, Product 034-6050H), nutrient broth (Unipath Ltd, CM1) and glucose, which has been found to be one of the best general purpose media for freeze-drying. A wide range of organisms, including some that do not survive in other freeze-drying media (e.g. photosynthetic and iron-oxidizing sulphur bacteria), have been successfully maintained using '*mist. desiccans*'. This suspending medium has been used for many years in the NCIMB and has proved to be a far superior suspending fluid for the more sensitive organisms than simple mixtures of serum and carbohydrate (including serum and inositol). Double-strength skimmed milk can be used as an acceptable alternative in most cases and is in fact used by some culture collections. However, it has been found that the sulphate-reducing group of bacteria show poor survival in this medium.

Attempts were made some years ago in the NCIMB to find a suspending medium more convenient to prepare than '*mist. desiccans*', in which experiments were carried out using sucrose–sodium glutamate mixtures. Initial survival levels were found to be high, but the viability of some organisms (notably the streptococci) fell drastically during storage. Sucrose–sodium glutamate mixtures are, therefore, not suitable for the long-term storage of freeze-dried bacteria.

B. Preparation of '*mist. desiccans*'

Horse serum, 100 ml (Life Technologies Ltd, Product No. 034-6050H), and nutrient broth, 33 ml (Unipath Ltd, CM1), are mixed in a 250 ml conical flask; 10 g glucose are then added slowly while shaking. Once all the glucose has dissolved, sterilization of the mixture is carried out using *pressure* filtration (either membrane or Seitz) rather than vacuum filtration (which produces troublesome frothing). The sterilized mixture is dispensed in 5 ml amounts in sterile screw-capped universal bottles and incubated at 30°C for 2–3 days as a sterility check. The '*mist. desiccans*' is stored at 20°C until used.

'*Mist. desiccans*', 5 ml, is used to harvest the growth from three universal bottle agar slope cultures. This will give an adequately dense suspension which can be dispensed in 0.1 ml amounts into approximately 25–30 ampoules.

The rest of the freeze-drying procedure and revival are essentially the same as described by Rudge (this volume, ch. 4.1). Storage of ampoules in the NCIMB is at 2°C; freeze-dried seed stocks are maintained at −20°C.

III. L-DRYING

Liquid drying (L-drying; Annear, 1958) is a useful alternative method of vacuum-drying for the preservation of bacteria that are particularly sensitive to the initial freezing stage of the normal lyophilization process. The intrinsic feature of this process is that cultures are prevented from freezing; drying occurs direct from the liquid phase.

A. Method

Sterile ampoules are prepared with numbered and dated filter paper strips, together with cotton-wool plugs as described by Rudge (this volume, ch. 4.1). A dense suspension of the bacterium in '*mist. desiccans*' is prepared and 3 drops (*c.* 0.1 ml) are dispensed into the ampoules by means of a sterile pipette, taking care not to leave any suspension on the sides of the ampoule. The cotton-wool plugs are trimmed and pushed down into the ampoule by means of a ramrod. The ampoules are attached vertically to the underside of a horizontal manifold, clamped above a glass tank containing water at 20°C, so that the ampoules can be immersed in the water to a depth of 40–50 mm. The manifold is connected via a diaphragm valve and phosphorus pentoxide trap to a rotary pump (Fig. 1).

With the valve closed the pump is switched on. The valve is then briefly opened wide (for approximately 0.5 s) and quickly closed again. This procedure removes most of the air from the system without causing violent removal of dissolved air from the

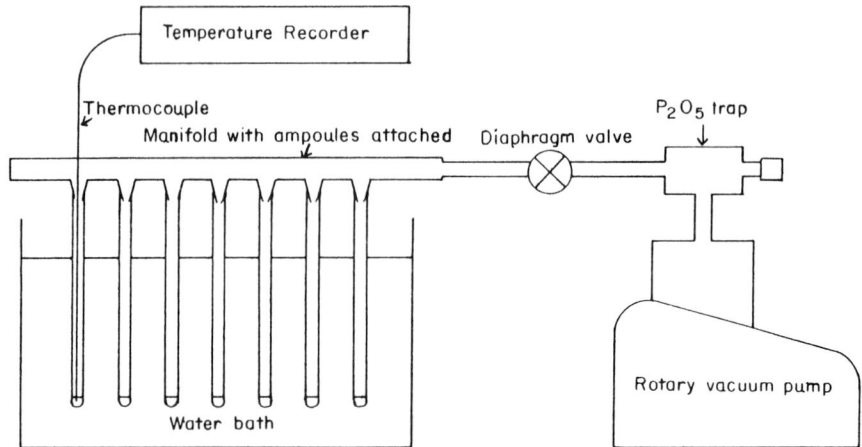

Fig. 1   Diagram of L-drying apparatus.

suspensions. The valve is then opened very gradually until the ampoule contents begin to de-gas (i.e. bubble). The rate of de-gassing is controlled by careful manipulation of the valve to prevent violent bubbling in the ampoules. In the event of this occurring the valve is closed quickly, to allow the bubbles to collapse, and then reopened cautiously. Violent de-gassing is undesirable since the culture would be dispersed up the sides of the ampoule; not only would this detract from the appearance of the ampoule, but it would also lead to charring of the material at the constriction and sealing stages.

Once de-gassing is complete, which takes approximately 5 min, the valve is opened fully to allow drying to take place. Adherence to this procedure will prevent freezing of the suspension.

After about 30 min. the contents of the ampoule will appear dry. The ampoules are then removed from the manifold, constricted, transferred to a secondary drier, dried overnight and sealed as described previously by Rudge (this volume, ch. 4.1). Resuscitation of L-dried cultures is as for freeze-dried cultures.

B. Shelf life

In the NCIMB bacteria such as spirilla and *Azomonas insignis* have been preserved by L-drying. These organisms are particularly sensitive to freeze-drying, but L-dried cultures have survived with good recovery levels for up to fifteen years. L-drying can therefore be considered as a suitable alternative to freeze-drying for bacteria that are susceptible to damage by freeze-drying.

IV. MICRODRYING

Microdrying is a modification of the freeze-drying method used in the NCIMB. It has several advantages over centrifugal freeze-drying since fewer manipulations are involved, resuscitation requires less skill and there is less risk of contamination as the resuspending stage is omitted.

A. Method

Ampoules are prepared exactly as for freeze-drying except that the filter strip used is a thick grade (Genzyme Biochemicals Ltd, No. 17). Bacterial suspensions, which should be as dense as possible, are prepared in '*mist. desiccans*'. Three drops (*c.* 0.1 ml) are dispensed, aseptically, on to the filter paper. The ampoules and the pipette should be held almost horizontally to ensure that the suspension falls only onto the filter paper and not onto the wall of the ampoule. All the suspension should be absorbed by the filter paper so that, although it appears damp, there is no excess liquid in the ampoule. Cultures are then freeze-dried in the usual way except that centrifugation and the replacement of cotton wool plugs by lint caps is not necessary.

Microdrying can be used as a convenient method of ensuring that known numbers of cells are placed in each ampoule. In this instance, sterile filter paper strips are placed aseptically in a sterile Petri dish, and a known aliquot (usually 0.1 ml) of a suspension containing a known cell concentration is added carefully to each. After all the suspension has soaked into the filter paper strips, they are transferred aseptically to sterile ampoules and a cotton wool plug inserted if required. Freeze-drying is carried out as previously described. After drying a further viable count is performed. The filter paper strips can then be used as standard inocula.

Resuscitation of microdried cultures is straightforward and does not require any manipulation with pipettes and resuspending medium. The ampoule is simply shaken or tapped to ensure that the filter paper strip is loose, opened, the cotton plug removed and the filter paper strip tipped aseptically into a broth or onto a slope of the appropriate medium.

B. Shelf life

Microdrying has been used in the NCIMB for about 12 years. Ampoules of various bacteria prepared in 1978 have recently been opened to test the survival of the cultures. Both Gram-negative and Gram-positive bacteria were tested. The results obtained indicate that, in general, the long-term survival and recovery of microdried cultures are comparable to those obtained with conventionally freeze-dried cultures.

V. FREEZING OVER LIQUID NITROGEN

Freezing over liquid nitrogen can be used for the preservation of a wide range of microorganisms. At the NCIMB this method is used for bacteria which do not survive freeze-drying or L-drying, for patent deposits, sensitive mutants, genetically manipulated strains and all bacteriophages. The latter are preserved by this method not because they cannot be freeze-dried or L-dried (although some cannot), but because of the risk of contaminating equipment, and consequently other cultures, with phage particles.

A. Bacteria

1. Method

Polypropylene cryotubes (2 ml, Life Technologies Ltd) containing 25–30 3 mm glass beads (available from craft shops) are individually wrapped in greaseproof paper, sterilized at 121°C for 15 min and dried at 50°C.

Bacteria are grown up as for freeze-drying and a thick suspension is made in an appropriate medium containing 10% (v/v) sterile glycerol. The glass beads in the cryotube are completely immersed in this suspension. The contents of the tube are agitated gently, to ensure that the beads are thoroughly coated, including the hole, with suspension and the excess suspension removed. The tubes are stored in aluminium racks in LR40 liquid nitrogen refrigerators (Union Carbide). All tubes are stored above the surface of the liquid nitrogen in the vapour phase to prevent the possible seepage of the liquid nitrogen into the tubes, thus eliminating contamination risks and safety problems on removal of the tubes.

Resuscitation of cultures is performed by removal of the glass bead from the frozen mass with sterile forceps and dropping it into the appropriate medium. The tubes should be kept frozen during this operation and are therefore transported to the laboratory over liquid nitrogen in a dewar flask, and only removed for a few seconds to facilitate the removal of the bead.

2. Shelf life

The glass bead method has been used in the NCIMB for approximately 10 years. To date, no problems have been encountered with most genera, although certain obligate methylotrophic bacteria which have been found to lose viability on storage over liquid nitrogen. Survival and recovery rates are comparable with conventional freeze-drying. The method is used as alternative to freeze-drying for susceptible bacteria, for example plasmid-containing strains. Compare also with the glass bead method described in Chapter 4.2.

B. Bacteriophages

1. Method   High-titre phage lysates are prepared using media and methods appropriate to the phage being preserved. Host cells and debris are removed by low-speed centrifugation followed by membrane filtration (0.45 μm pore size). The cell free lysates are dispensed in 1 ml aliquots in 2 ml sterile cryotubes (Life Technologies Ltd) and stored over liquid nitrogen as described above for bacteria.

2. Shelf life   Most coliphages, after an initial loss of titre of approximately one order of magnitude due to the freezing process, will survive for at least 10 years without further significant loss of titre. An exception is coliphage MS2 which does not survive for more than 1.5 years. *Pseudomonas aeruginosa* phages have been shown to survive for at least 10 years and *Staphylococcus aureus* phages for 9 years.

Marine phages tend to have shorter shelf lives. For example, *Pseudomonas* PM2 phage often does not survive for more than 2–3 years.

4.5. LOW-TEMPERATURE FREEZING OF MICROORGANISMS ON SILICA GEL

T. M. SIDYAKINA

All-Union Collection of Microorganisms, Institute of Biochemistry
and Physiology of Microorganisms,
USSR Academy of Sciences, Pushchino, Moscow Region, USSR

I. Introduction 65
II. Details of method 66
   1. Preparation of suspending medium 67
   2. Distribution of suspension onto desiccated
      silica gel 67
   3. Preliminary drying and cooling in a
      refrigerator 68
   4. Distribution of granules to plastic tubes 68
   5. Freezing tubes containing microbial cells on
      carriers....................................... 68
   6. Checking the product 68
III. Organisms successfully preserved 68
IV. Advantage of method 70

I. INTRODUCTION

The All-Union Collection of Microorganisms (VKM) is the major microbial resource centre in the USSR. It maintains a wide range of microorganisms (about 10 000 strains of actinomycetes, bacteria, fungi, yeasts and genetically marked strains with re-combinant DNA and plasmids) and carries out a substantial research program on investigations into viability, maintenance of biochemical activity and genetic stability of microorganisms of various taxonomic groups, especially biotechnologically impor-tant strains, after preservation by different methods. Details of methods used at the VKM are in three books (Sidyakina, 1985, 1988, 1989).

Freeze-drying or drying under vacuum from the frozen state

MAINTENANCE OF MICROORGANISMS 2nd Edn
ISBN 012–410351–0

has been successfully used for a long time for storing microorganisms. Nevertheless, many non-sporulating microbial cultures possess little tolerance to freeze-drying procedures and have low viability when preserved by this method, their viability declining in storage even at a temperature of 4°C.

Cryopreservation or storage at the temperature of liquid nitrogen (-196°C) came into use for preservation of microorganisms more recently. With cryopreservation, the relative number of viable cells is substantially higher than with freeze-drying, and the viability is retained during long-term storage in liquid nitrogen. For this reason, the possible period of time for which microorganisms can be stored is greatly increased. Nevertheless, to successfully cryopreserve most non-sporulating microorganisms (especially eukaryotes — fungi and yeasts), preliminary programmed freezing with selection of the most effective cryoprotectants and optimal cooling rates is necessary.

To preserve different microbial species by deep freezing and cryopreservation, a pre-drying of cells on desiccated carriers such as silica gel, glass beads, polymeric materials, filter paper strips and other materials frequently leads to improved survival levels (Sidyakina, 1985). Immobilized microbial cells usually remain viable and preserve their biochemical activity under the action of the extreme conditions associated with conservation far better than free cells. It is also generally the case that the more a microbial cell is dehydrated, the better it sustains freezing and repeated freeze-thawing.

VKM has developed and successfully used a simple technique for deep-freezing microbial cells on desiccated silica gel (USSR Author's Certificate No. 1442544, 1988). This method has been successfully used for a number of years to preserve bacteria, fungi, yeasts, actinomycetes and genetically marked strains carrying recombinant DNAs and plasmids (Sidyakina, 1988).

II. DETAILS OF METHOD

Figure 1 illustrates the process schematically. The steps involved are:

(1) preparation of suspending medium;
(2) distribution of suspension to desiccated silica gel;
(3) preliminary drying and cooling in refrigerator;
(4) distribution of granules with cells to plastic tubes;
(5) freezing of tubes with microbial cells on carriers;
(6) checking of product.

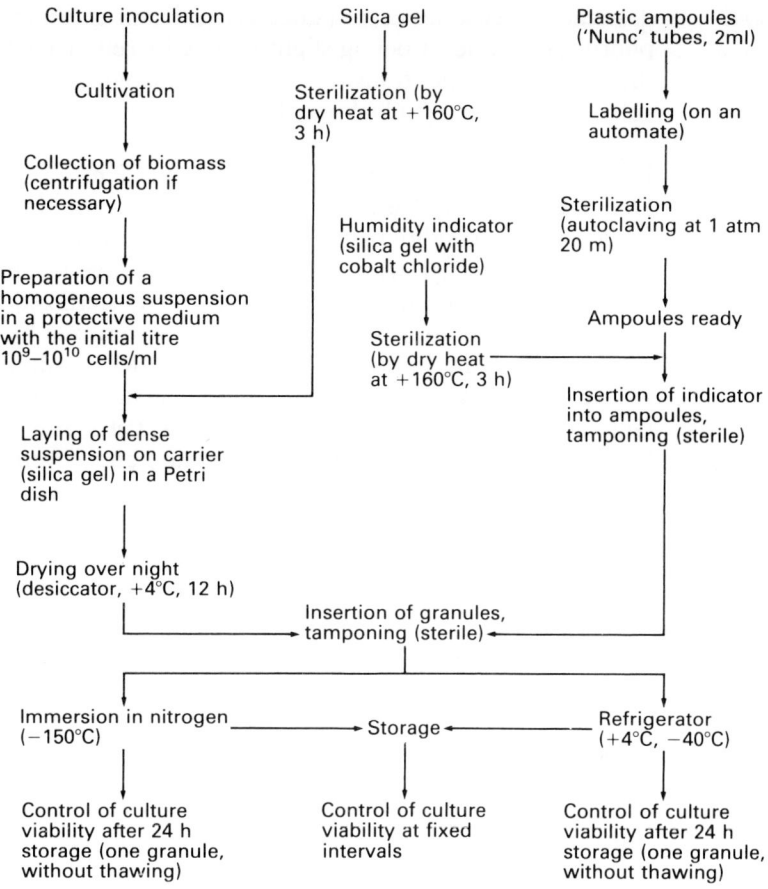

Fig.1   Low-temperature freezing of microorganisms on desiccated silica gel.

1. Preparation of suspending medium

The microbial cells grown under optimal condition until the stationary growth phase (until abundant sporulation occurs, in the case of spore-forming microorganisms) are suspended in 10% (v/v) glycerol solution and/or in 10% skimmed milk to obtain a homogeneous dense suspension containing 10^9 to 10^{10} cells ml^{-1}. The suspension is cooled at 4°C for 4 h.

2. Distribution of suspension onto desiccated silica gel

Granules (3 mm) of porous silica gel KSM-6 (GOST 3956-76) are used as carriers. The sterilization of the silica gel is carried out using dry heat at 160°C for 3 h. Under aseptic conditions, the suspension is applied with a Pasteur pipette to previously desiccated (during sterilization) sterile silica gel granules placed in a Petri dish. The process is heat-releasing, hence a preliminary cooling of the suspension is recommended.

3. Preliminary drying and cooling in a refrigerator

Granules with cells are dried overnight (12 h) at 4°C in a desiccator over a desiccant (silica gel, calcium chloride or phosphorus pentoxide). Cooling slightly impedes cell metabolism and thus promotes the transition to cold anabiosis, as with the subsequent drying (Sidyakina, 1988).

4. Distribution of granules to plastic tubes

Polypropylene ampoules (e.g. cryotubes 2 ml, A/S Nunc, Denmark) are sterilized by autoclaving at 1 atm for 20 min. Under aseptic conditions, several grains of calcinated moisture indicator (silica gel with cobalt chloride, 'Merck', FRG) are placed with sterile forceps into a 2 ml plastic sterilized ampoule with a screw-cap. Moisture indicator can be sterilized by dry heat at 160°C for 3 h. Indicator (bright blue after sterilization and drying) is covered with a thin layer of sterile cotton wool. Granules of inoculated silica gel from the Petri dish are transferred into ampoules (20–25 per ampoule) and covered with a small flock of cotton wool; the caps are then screwed tightly.

5. Freezing tubes containing microbial cells on carriers

Ampoules containing the microbial cultures are stored at −150°C in the vapour phase of nitrogen in nitrogen refrigerators or at −40°C, or −70°C in deep freezers. Duplicate cultures (for distribution) are stored in a refrigerator at 4°C or in the freezing chamber at −12°C. Storage of microorganisms at temperatures in the range of −70°C to −130°C is now possible in many laboratories because of the ready availability of commercial deep freezers operating within this range. Many collections of microorganisms use cryopreservation and have low-temperature banks utilizing liquid nitrogen. The lower the storage temperature, the better the viability of cultures preserved by this method and the longer the shelf life.

6. Checking the product

After 1–2 days of storage, the viabilities of cultures are checked. The granules of silica gel are withdrawn from the ampoules (one granule from each ampoule) and are used as inocula on Petri dishes, with appropriate media. The plates are incubated under optimal conditions. Ampoules containing the remaining granules of silica gel are replaced for storage. To prevent thawing during sampling, ampoules are transferred from the low-temperature bank to the laboratory bench in a paraffin wax beaker cooled with liquid nitrogen.

III.   ORGANISMS SUCCESSFULLY PRESERVED

This method was developed for long-term storage of genetically marked strains of bacteria having recombinant plasmids, and was later successfully used in VKM to preserve a wide range of bacteria, actinomycetes, fungi, yeasts and recombinant clones. Some of the recombinant clones, plasmids and strains, which were successfully preserved on silica gel following storage at −70°C and −150°C during 6 years of investigation (to date) are shown in Table I.

Table I: Recombinant clones, plasmids and strains under investigation by various preservation methods

Clone	*E. coli* strain	Plasmid	Cloned fragment	Antibiotic marker
pND10	C 600	pBR325	calico salmon insulin gene	T_c
pND77	C 600	pBR325	calico salmon insulin gene	A_p
pHS35	HB 101	pKN001	Alu — family, DNA repeat	T_c
pEcHD01	HB 101	pBR322	Anonymous sequence	A_p
pBR13	HB 101	pBR325	Human DNA, *Eco*RI repeats	T_c
pARI25	HB 101	pBR325	Human DNA, *Eco*RI repeats	T_c

The influence of different methods of preservation (freeze-drying and storage at 4°C, low-temperature freezing on silica gel with storage at −70°C and −150°C, cryopreservation with storage at −196°C) on the survival and genetic stability of *Escherichia coli* HB 101, containing a recombinant plasmid with a human DNA fragment, was investigated. The genetic stability was estimated by following the maintenance of selective markers of antibiotic resistance and the stability of restriction sites in recombinant DNA. It was shown that the low-temperature freezing on desiccated silica gel and cryopreservation are the optimal methods for preservation, giving high survival level and genetic stability. *Escherichia coli* strains carrying recombinant DNAs and plasmids maintained the selective markers of antibiotic resistance and the restriction sites. The strains remained viable and preserved their genetic stability over 6 years of examination (Sidyakina and Golimbet, in press).

The method described can be recommended for the reliable long-term storage of microorganisms and recombinant clones without loss of marker properties.

IV. ADVANTAGES

A substantial advantage of low temperature freezing of microbes on desiccated silica gel is the absence of genetic changes in microbial cells, which undergo less damage compared to that caused by drying and lyophilization. Some microbial species unable to survive freeze-drying and ultra-rapid cryopreservation techniques from cell suspensions successfully survive storage on silica gel in the vapour phase of nitrogen at $-150°C$. This method therefore broadens the range of cultures surviving long-term storage. Other advantages of this method are its simplicity and convenience, minimal pre-treatment, quick recovery of frozen cells, the possibility of withdrawing a small part of the material and replacing the remainder for further storage, fewer biohazards and greater safety for personnel. The method is highly practicable since it reduces the number of ampoules being stored, thus increasing the efficient use of storage space.

4.6. MAINTENANCE OF ANAEROBIC BACTERIA

C. S. IMPEY and B. A. PHILLIPS[1]

Cherry Valley Farms Ltd,
North Kelsey Moor, Lincoln, UK
[1] National Collection of Food Bacteria,
AFRC Institute of Food Research, Reading, UK

I. Introduction 71
II. Culture methods for anaerobes 72
III. Media 73
    A. Media for use with the Hungate technique 73
    B. Media for use in screw-cap McCartney bottles 75
IV. Maintenance by subculture 76
    A. Cultures requiring the Hungate technique 76
    B. Cultures in screw-cap bottles 76
V. Maintenance by freezing 76
VI. Maintenance by freeze-drying 76
    A. Apparatus and equipment 76
    B. Preparation of the suspension for freeze-drying . . . 77
    C. Freeze-drying method 78
    D. Opening ampoules 79
    E. Quality control 79
    F. Survival on drying 79
    G. Shelf-life 80
VII. General notes 80

I. INTRODUCTION

Although media and methods for the isolation and culturing of anaerobes have improved greatly over the past few years, the preservation of cultures still causes problems. Maintenance by periodic subculture is inconvenient, time-consuming and often results in contamination or loss of viability.

When cultures are freeze-dried the resulting ampoules, if correctly stored, will remain viable for periods far beyond any practical requirement. In addition, the freeze-dried culture is the only practical form in which strains can be exchanged

MAINTENANCE OF MICROORGANISMS 2nd Edn
ISBN 012–410351–0

between researchers. Other maintenance methods are described in Chapters 4.7 and 4.10.

II. CULTURE METHODS FOR ANAEROBES

Current methods for the cultivation of anaerobes may be conveniently divided into those in which the media are prepared and used under an oxygen-free atmosphere, and those in which the media are prepared and inoculated in air and which may or may not be incubated with an oxygen-free headspace.

For strict anaerobes the method first described by Hungate (1950) may be used. All manipulations, including media preparation, are carried out under a constant flow of carbon dioxide or nitrogen which has been passed over heated copper to remove any remaining traces of oxygen. Further details of this method can be found in the publications of Hungate (1966, 1969), Latham and Sharpe (1971), Barnes and Impey (1974) and Holdeman *et al.* (1977). Anaerobic cabinets also maintain a constant oxygen-free headspace over the media and are now available from manufacturers of scientific equipment (Don Whitley Scientific Ltd; Raven Scientific Ltd).

However, many anaerobes can be grown without the use of the Hungate technique or anaerobic chambers provided conditions are properly controlled. This is especially true of clinical isolates and evidence now suggests that pathogenic anaerobes are among the more oxygen tolerant (Rosenblatt *et al.*, 1973). Broth media containing reducing agents such as cysteine, together with small amounts of agar to minimize diffusion of oxygen, are used in 1 oz McCartney bottles with metal screw-caps. The bottles are filled with about 20 ml of broth to leave a minimal headspace and immediately prior to inoculation are held in a boiling water bath for 20 min to expel oxygen. After inoculation, with *c.* 0.25 ml of culture, the caps are screwed down tightly to prevent access of oxygen and the broths are then incubated at 37°C for 1–2 days. Agar plates should be pre-reduced in anaerobic jars for at least 24 h before inoculation and then returned to the jars for incubation. The gas mixture in the anaerobic jars is 80% N_2 + 10% H_2 + 10% CO_2; alternatively, gas-generating sachets (Unipath Ltd; B-D Laboratory Products) can be used.

III.   MEDIA

A. Media for use with the Hungate technique

		Per litre
1. MM10	Cellobiose	1.0 g
broth modified	Maltose	1.0 g
from M10	Glucose	1.0 g
(Caldwell	Starch (soluble)	1.0 g
and Bryant,	Yeast extract, B127 (Difco)	2.0 g
1966)	Trypticase, BBL11921 (B-D)	2.0 g
	Mineral solution I	75 ml
	Mineral solution II	75 ml
	Haemin solution	10 ml
	VFA mixture	3.1 ml
	Resazurin solution	1.0 ml

pH 6.8

Separate additions: 20 ml l^{-1} of 2.5% (w/v) L-cysteine hydrochloride
50 ml l^{-1} of 8% (w/v) sodium carbonate

Preparation: All the ingredients except cysteine and carbonate are
dissolved in glass-distilled water, the pH is adjusted to 6.8 with
1 M NaOH and made up to 93% of final volume. This is sterilized
in a cotton wool plugged flask by autoclaving at 121°C for
15 min. The 2.5% (w/v) cysteine solution is prepared in a screw-
cap McCartney bottle with a small headspace and the 8% (w/v)
sodium carbonate solution in a small cotton wool plugged flask
with a large headspace; both are autoclaved at 121°C for 15 min.
While the flasks are still hot from the autoclave, short, sterile
gassing jets are inserted and oxygen-free carbon dioxide is passed
over the surface of the medium and of the carbonate. When the
medium has cooled to below 50°C the cysteine solution is added,
followed by the carbonate; long gassing jets are inserted below
the liquid surface and bubbled for at least 30 min. The complete
medium is distributed into sterile stoppered tubes using standard
Hungate technique, and incubated at least overnight to check for
contamination and for oxidized (pink) tubes. This method of
preparation is that of Latham and Sharpe (1971) but an alternative method for preparing the medium is described by Barnes and
Impey (1974).

(a) *MM10 medium solutions and mixtures*
 (i) Mineral solution I:
   K_2HPO_4 6.0 g l^{-1}
   in glass distilled water; store at 4°C.

 (ii) Mineral solution II

	g l^{-1}
NaCl	1.2
$(NH_4)_2SO_4$	12.0
KH_2PO_4	6.0
$CaCl_2$ (anhydrous)	1.2
$MgSO_4$ $7H_2O$	2.5

   in glass distilled water; store at 4°C.

 (iii) Haemin solution
   10 mg of haemin is dissolved in 1 ml 1 M NaOH and the volume made up to 100 ml with glass distilled water; store at 4°C.

 (iv) VFA Mixture

Acetic acid	17 ml
Propionic acid	6 ml
n-Butyric acid	4 ml
iso-Butyric acid	1 ml
DL-α-methyl-*n*-butyric acid	1 ml
n-Valeric acid	1 ml
iso-Valeric acid	1 ml
Phenyl-acetic acid	1 g

   Mix well and store at 4°C.

 (v) Resazurin solution.
   0.1% (w/v) resazurin in sterile glass distilled water. Store at 4°C.

(b) *Additions and supplements for MM10*
 (i) Agar—for agar slants 1.5% (w/v) agar (Difco) is added to the broth medium.

 (ii) Lactate—the growth of *Veillonella* and *Megasphaera* species is enhanced by the addition of sodium lactate to a final concentration of 1% (w/v).

 (iii) Rumen fluid—although this medium was developed to obviate the need for rumen fluid, some organisms such as *Eubacterium cellulosolvens* will grow much better if clarified rumen fluid (Bryant and Robinson, 1961) is added to 15% (v/v) final concentration.

(iv) Faecal extract—Barnes and Impey (1978) found that a significant number of anaerobes of poultry origin had a requirement for an aqueous extract of chicken faeces. The extract is prepared by autoclaving at 121°C for 30 min equal quantities (w/w) of chicken faeces and water. The sludge is centrifuged (1500 g for 20 min) and, after decanting, the supernatant is left at 1°C overnight. The supernatant is further centrifuged (16 000 g for 20 min) to remove remaining debris and the pH is adjusted to 7.0–7.2. It is stored sterile (autoclave at 121°C for 15 min) and when required is incorporated in the medium at a final concentration of 10% (v/v). Extracts of faecal material from other sources can be prepared and used in the same manner.

(v) Liver extract—Barnes and Impey (1978) also found anaerobes with a requirement for liver extract. This extract is prepared by dissolving 27 g of dehydrated liver, B133 (Difco Laboratories Ltd), in 200 ml glassdistilled water, heating to 50°C and holding at this temperature for 1 h. The mixture is then boiled, cooled and centrifuged (1500 g for 20 min) and the pH of the supernatant adjusted to 7.0–7.2. The extract is stored sterile (autoclave at 121°C for 15 min) and used in the medium at a final concentration of 5% (v/v).

B. Media for use in screw-cap McCartney bottles

1. VL broth (Barnes and Impey, 1971; modified from Beerens et al., 1963) This contains (g l^{-1}): tryptone, L42 (Unipath Ltd), 10; NaCl, 5; beef extract, Lab-Lemco powder L29 (Unipath Ltd), 2.4; yeast extract, B127 (Difco Laboratories Ltd), 5; cysteine hydrochloride, 0.4; glucose, 2.5; agar (New Zealand), 0.6; pH 7.2–7.4. The medium is sterilized by autoclaving at 121°C for 15 min in 19 ml amounts in 1 oz McCartney bottles with metal caps.

2. BGP broth This is VL broth with the glucose level reduced to 1 g l^{-1} and with the addition of Na_2HPO_4 at 4 g l^{-1}.

3. VLhlf and BGPhlf These media are VL and BGP broths containing 1 μg ml^{-1} haemin (see above), 5% liver extract (see above) and 5% chicken faecal extract (see above).

4. Cooked meat broth This can be made using either of the formulations described by Cowan (1974) or by using meat granules, Lab 24 (London Analytical & Bacteriological Media Ltd) in nutrient broth. The medium is distributed in 20 ml amounts in 1 oz screw-cap bottles and sterilized by autoclaving at 120°C for 15 min. It is recommended that the depth of meat granules should be about 5 cm.

IV.   MAINTENANCE BY SUBCULTURE

A. Cultures requiring the Hungate technique

MM10 agar with supplements as required is used in tubes that are sloped with a deep butt. The medium is inoculated by stabbing into the butt. Cultures can be kept in this medium for 1–4 weeks and are stored at room temperature in the dark.

B. Cultures in screw-cap bottles

Anaerobes may be kept for 1–4 weeks in low-glucose buffered media such as BGP or BGPhlf. The broths should be stored at room temperature in the dark with the caps screwed down tightly and only opened for use once. Non-sporing anaerobes are very sensitive to acid conditions and the use of unbuffered glucose-containing media such as VL is not recommended.

Media suitable for non-sporing anaerobes may also be used for clostridia, but cooked meat broth is often more satisfactory. The more putrefactive readily sporing clostridia will survive in this medium for several years, but the poorer spore formers for only 6–12 months.

V.   MAINTENANCE BY FREEZING

Anaerobes that grow well on MM10 agar slopes can be stored at −80°C for about 3 months. There are, however, a number of disadvantages with this method of preservation: the tube size limits the numbers that can be stored; the bungs used in the Hungate tubes harden in the cold and may become dislodged; and slow thawing may well damage cultures irreparably.

VI.   MAINTENANCE BY FREEZE-DRYING

A. Apparatus and equipment

1. Freeze-drying machines   Any centrifugal freeze-dryer is suitable for the preservation of anaerobic bacteria. The methods described here have been used in conjunction with freeze-dryer models Modulyo 4K, 5PS,

EF03 and EF6 (Edwards High Vacuum International).

2. Ampoules   Before use neutral glass ampoules (Anchor Glass Co. Ltd) are washed in detergent and thoroughly rinsed, first with tap water and then demineralized water. They can be labelled by inserting strips of Whatman No. 1 filter paper (Scientific Supplies Co. Ltd) with the culture number and date of drying written either in pencil or non-toxic ink (Lapage *et al.*, 1970).

The ampoules are capped with 10 × 50 mm glass specimen tubes (Scientific Supplies Co. Ltd) or non-absorbent cottonwool plugs and dry-sterilized.

3. Primary drying caps   Caps for primary drying are made using surgical lint sewn at the edge. The overall size is approximately 20 × 40 mm; the cap should be an easy fit on the ampoule, but tight enough to withstand handling and spinning in the dryer centrifuge. The caps are sterilized in glass Petri dishes.

4. Secondary drying plugs   Plugs for secondary drying are made by covering butter muslin with a thin (3–5 mm) layer of non-absorbent cotton wool and cutting into 12 mm squares. The squares are packed in glass Petri dishes, muslin side up (layers interleaved with paper), and dry-sterilized. Alternatively, plugs from sterile ampoules can be used.

5. Pipettes   Both standard 1 ml pipettes and 30-dropper Pasteur pipettes (Harshaw Chemicals Ltd; John Poulten Ltd) are plugged with non-absorbent cotton wool and dry-sterilized in cans. It is important that the 30-dropper pipettes are long enough to reach the bottom of the ampoules (*c.* 100 mm).

B. Preparation of the suspension for freeze-drying

1. Growth media   Organisms for freeze-drying must be grown in a medium giving rapid and vigorous growth. For strains grown using the Hungate technique, MM10 broth should be used and VL broth for screw-cap bottle cultures. Both media have supplements of rumen fluid, faecal extract, liver extract, etc., added as required.

2. The suspension   Hungate cultures are centrifuged (2500 *g* for 15 min), the supernatant poured off, and 1 ml of suspending fluid added to the centrifuged deposit. The suspending fluid is 7.5% glucose in horse serum (Tissue Culture Services Ltd), sterilized by filtration (Phillips *et al.*, 1975). Up to the point where the supernatant is poured off, Hungate cultures should be kept under a carbon dioxide headspace, but once preparation of the suspension is complete this can be dispensed with provided there is no undue delay before freeze-drying is started.

Cultures in screw-cap bottles are centrifuged (2500 g for 15 min) and after the supernatant has been poured off 1 ml of 16% glucose, which has been held in a boiling water bath for 10 min and cooled just before use, is added to the centrifuged deposit.

C. Freeze-drying method

1. Primary drying

The bacterial suspension is distributed in approximately 0.1 ml quantities per ampoule, the glass cap/cotton wool plug is replaced by a lint cap and the ampoules placed in the dryer centrifuge. The time taken from filling the first ampoule to starting the freeze-dryer should be kept as short as possible— about 30 min for cultures grown in screw-cap bottles and 15 min for Hungate cultures. The centrifuge should be run for up to 10 min and primary drying continued for 3 h.

Freeze-dryers using phosphorus pentoxide, methanol/solid CO_2 cold fingers and mechanically frozen trays have all proved successful for drying anaerobes.

2. Ampoule constriction and secondary drying

The ampoules are removed from the primary drying chamber and the lint caps replaced with plugs. Using sterile pointed forceps the muslin/cotton wool plugs are held in the centre and then pushed into the ampoule neck. If plugs from sterile ampoules are used they should be carefully transferred to the ampoule containing the dried material and then trimmed off to the required length (*c*. 10 mm). The plugs are pushed down the ampoule, using a rod fitted with a stop, so that the bottom of the plug is approximately 10 mm above the top of the dried material. The pointed forceps and the plug push rod should be sterilized by dipping in 70% alcohol and flaming between batches of ampoules of any one strain.

The ampoules are then constricted using either a constricting machine (Edwards High Vacuum International) or an oxy-gas torch (Jencons (Scientific) Ltd) with a fine jet. Care should be taken to make the bore of the constriction fairly fine but with thick walls, so that when the ampoules are finally sealed off a good strong point is formed.

Fit the ampoules onto the secondary drying head, fitting any unused nipples with empty ampoules. Where phosphorus pentoxide was used as the primary drying desiccant the trays are topped up with fresh material for secondary drying. With methanol/CO_2 cold fingers and mechanically refrigerated traps, phosphorus pentoxide in trays is also used as a secondary drying desiccant. Secondary drying (under vacuum) is continued overnight (15–20 h).

The ampoules are then sealed under vacuum using a Flaminaire torch (Longs Ltd) and tested for vacuum using a high-frequency vacuum tester (Edwards High Vacuum International). The discharge should be a blue-white colour; if the colour is mauve the ampoule should be left for a few hours and then retested. No discharge means no vacuum.

D. Opening ampoules

The ampoule is marked with a glass knife at the level of the centre of the cotton-wool plug. The area of the mark is wiped with 70% alcohol, flamed, and then cracked by applying a fine, red-hot rod. Approximately 30 s is allowed for air to enter the ampoule, then the pointed end of the ampoule is removed. The cotton-wool plug is replaced with a fresh one from a sterile ampoule. The dried culture material is reconstituted using suitable broth media; the medium is usually the same as that used to grow the organism for drying.

E. Quality control

1. Viability   To check that the organism is still viable, one ampoule of each batch of a strain should be opened using a suitable broth.

2. Counting   In general, a simple growth or no-growth check for viability will be sufficient but, if an indication of survival rate is required, some counting method must be used. Surface plating or roll tubes can be used, but a much more practical method is to prepare tenfold dilutions to extinction using appropriate broth media (Phillips *et al.*, 1975).

3. Purity   Anaerobes, especially those cultivated by the Hungate technique, are very prone to contamination. When the test for viability is carried out the culture should also be checked for contaminating aerobes and anaerobes. It is also useful to check the purity of the suspension used for drying.

F. Survival on drying

Most anaerobes seem to be able to survive freeze-drying, but the loss in viable count will vary considerably. The variation is not dependent on genus; even different strains of the same species show varying survival levels. Also, the degree of anaerobiosis required to culture an organism is no index of its ability to survive freeze-drying; some of the strictest anaerobes show comparatively small losses on drying.

G. Shelf life

Although the initial loss on drying may be high, if the ampoules are stored in the dark at a cool (2–10°C) even temperature, further losses are low. The viability of anaerobes preserved in this way has remained high for at least 10 years, and even after 22 years cultures are still alive.

VII.   GENERAL NOTES

(1) Rather than use a multiplicity of media for drying anaerobes it is often possible to use one medium for cultures in screw-cap bottles and one for cultures requiring the use of the Hungate technique, provided that the media selected will grow the most nutritionally demanding of the organisms being handled.

(2) Freeze-dryers require little maintenance, apart from the need to change vacuum pump oil and clean spilled phosphorous pentoxide from desiccant chambers. However, there are a number of seals that are made and broken each time the dryer is used and these will require replacement at fairly frequent intervals if the dryer is to maintain its performance. These seals include the secondary dryer nipples and the 'O' and 'L' rings of the desiccant chambers and centrifuge chamber bell jars.

(3) In order to minimize the exposure of cultures to oxygen, the freeze-dryer should be prepared with all desiccant trays filled or mechanical freezers run down to the required temperature, and the general vacuum tightness of the dryer checked before any suspensions are prepared for distribution into ampoules.

4.7. MAINTENANCE OF PHOTOTROPHIC BACTERIA

KHURSHEED AHMAD MALIK

DSM-Deutsche Sammlung von Mikroorganismen und
Zellkulturen GmbH, Braunschweig, FRG

I.	Introduction	81
II.	Culture methods and media	83
	1. Growth medium for Rhodospirillaceae	84
	2. Modified Pfennig's medium for purple sulphur bacteria	85
	3. Modified Pfennig's medium for green sulphur bacteria	87
	4. Modified growth medium for *Chloroflexus*	87
III.	Maintenance by subculturing	88
IV.	Maintenance by freezing	89
	1. Equipment	90
	2. Preparation of cell suspension for freezing	91
	3. Filling ampoules and freezing	92
	4. Revival of cultures	92
	5. Estimation of viability counts	92
V.	Maintenance by freeze-drying	93
	A. Freeze-drying under aerobic conditions	94
	B. Freeze-drying under anaerobic conditions	96
VI.	Maintenance by liquid-drying	98
VII.	General notes	98

I. INTRODUCTION

The anoxygenic phototrophic (photosynthetic) bacteria represent a group of predominantly aquatic bacteria. They are able to grow under anaerobic conditions by photosynthesis without oxygen production. The presence of bacteriochlorophylls and of carotenoid pigments is common to all species (Pfennig and Trüper, 1974, 1989). The photosynthetic metabolism differs from that of the cyanobacteria, algae and green plants in that water cannot serve as an electron donor. They photoassimilate carbon dioxide

MAINTENANCE OF MICROORGANISMS 2nd Edn
ISBN 012–410351–0

through the reductive pentose phosphate cycle or the reductive citric acid cycle. The photosynthetic CO_2 assimilation depends on the utilization of external electron donors, such as reduced sulphur compounds, molecular hydrogen or organic compounds.

When grown under optimal phototrophic conditions, the colours of cell suspensions of phototrophic bacteria range from purple-violet to purple, red, orange-brown, yellow-brown, brown or green. The diameter of individual Gram-negative cells ranges from 0.3 to >6 µm and they are spherical, spiral, rod or vibrioid-shaped; cells are unicellular or uniseriately multicellularly filamentous. Spores, cysts or other resistant structures, which could otherwise help survival under adverse conditions, are not known.

Phototrophic bacteria comprise two cytologically different groups:

(1) Purple bacteria (Chromatiaceae, Ectothiorhodospiriaceae) and purple nonsulphur bacteria (such as Rhodospirillaceae), in which the photosynthetic pigments are located in an intracytoplasmic membrane system that is continuous with the cytoplasmic membrane.
(2) Green bacteria (Chlorobiaceae, including the genera *Chloroherpeton*, *Chloroflexus*, *Heliothrix*, *Chloronema*) in which the photosynthetic pigments are located in the cytoplasmic membrane and in the chlorosomes that underlie and are attached to the cytoplasmic membrane (Imhoff *et al.*, 1984; Stackebrandt and Woese, 1984).

In addition to these, the obligatory anaerobic and phototrophic genus *Heliobacterium*, and the aerobic, chemoorganoheterotrophic genus *Erythrobacter* have been assigned to genera *incertae sedis* as these cannot be affiliated with the purple or green bacteria.

Difficulties in the cultivation and laboratory maintenance of phototrophic bacteria are well known (Van Niel, 1971; Pfennig and Trüper, 1981; Trüper and Imhoff, 1981; Bose, 1963; Malik, 1978, and later). Although media and methods for their isolation and culture have improved greatly over the past few years, their preservation and maintenance still causes problems (Pfennig and Trüper, 1981; Biebl and Pfennig, 1981; Malik, 1983). Generally, the phototrophic bacteria are maintained as living cultures. However, regular subculturing of large stocks is very time-consuming and is liable to cause mutation or contamination, especially when the cultures are maintained in a liquid selective media.

Several microorganisms can be preserved with good viability and stability by freezing in liquid nitrogen ($-196°C$), by freeze-drying (lyophilization) and by liquid-drying (Lapage *et al.*, 1970;

Malik, 1976; 1987a; Malik and Claus, 1987). However, the preservation of anaerobic phototrophic bacteria by such conventional (aerobic) methods usually results in low viability; the retrieval of strictly anaerobic Chlorobiaceae, Chromatiaceae and brown Rhodospirillaceae is especially complicated and very time-consuming. Some strains of anoxygenic phototrophic bacteria can grow as chemoheterotrophs and chemoautotrophs in the dark under microaerobic to aerobic conditions and are easier to maintain. Others are very sensitive to oxygen and can only grow as photoautotrophs or photoheterotrophs under anaerobic conditions in the light. In such cultures, the photopigment content (carotenoid, bacteriochlorophyll or porphyrin) is high and is responsible for their sensitivity to light and oxygen (Malik, 1984a). Thus failure to preserve such cultures by conventional aerobic methods is mainly due to their sensitivity to oxygen in the presence of light, presumably resulting in membrane damage due to photodynamic effects. During several years of experimentation, various preservation methods have been optimized to achieve successful maintenance of a large collection of anoxygenic phototrophic bacteria (Biebl and Malik, 1976; Malik, 1978, 1984b, 1987b, 1988b, 1989).

The procedures used at the Deutsche Sammlung von Mikroorganism und Zellkulturen (DSM) for maintaining phototrophic bacteria are described here.

II.   CULTURE METHODS AND MEDIA

Almost complete anaerobiosis is needed for media preparation and growth and maintenance of phototrophic bacteria. Simplified media and a convenient anaerobic method described for the cultivation of phototrophic bacteria in 1983 (Malik, 1983) proved most suitable for good phototrophic growth of almost all strains at the DSM — several species of Chromatiaceae, Chlorobiaceae, Chloroflexaceae, Ectothiorhodospiriaceae, Rhodo-spirillaceae and Heliobacterium. For strict anaerobes the principle of Hungate's (1950) technique are applied during media preparation and during inoculation or transfers.

For purple sulphur bacteria and green sulphur bacteria various media have been described. The media developed by Pfennig and Trüper (1981) have been used for many years for the successful cultivation of these bacteria. A modified medium based on this has been described earlier (Malik, 1983). The composition of the medium for phototrophic growth of Rhodospirillaceae (Biebl and Pfennig, 1981; Malik, 1983) and a simplified Pfennig's medium for the successful cultivation of a broad range of purple and green sulphur bacteria are described below.

1. Growth medium for Rhodospirill- aceae	Yeast extract	0.30 g
	Di-sodium succinate	1.00 g
	Ammonium acetate	0.50 g
	Ferric citrate solution (0.1% in H_2O)	5.00 ml
	KH_2PO_4	0.50 g
	$MgSO_4.7H_2O$	0.40 g
	NaCl	0.40 g
	NH_4Cl	0.40 g
	$CaCl_2.2H_2O$	0.05 g
	Vitamin B_{12} solution (10 mg in 100 ml H_2O)	0.40 ml
	Trace element solution SL-6 (see below)	1.00 ml
	Distilled water	1050 ml

The pH is adjusted to 6.8. The medium is boiled under a stream of N_2 gas for a few minutes and 45 ml of the medium is distributed into 50 ml screw-capped bottles (already flushed with N_2 gas) with a rubber septum. Each bottle is bubbled with N_2 gas, closed immediately and screwed tightly. Bottles are autoclaved at 121°C for 15 min. Sterile syringes are used to inoculate and remove samples. Incubation is done in the light using a tungsten lamp.

For brown and other oxygen-sensitive Rhodospirillaceae, 300 mg of L-cysteine (0.03% end concentration) are added to the boiling medium and the pH is readjusted to 6.8. Alternatively, to the prepared medium in bottles, neutralized sulphide solution is injected (0.005–0.01% end concentration). For *Rhodopseudomonas acidophila* and *Rhodomicrobium vannielii*, the pH is adjusted to 5.6. For *Rhodopseudomonas globiformis*, di-sodium succinate is omitted and, instead, 1.5 g mannitol and 0.5 g sodium gluconate are added and pH is adjusted to 4.9. To 45 ml sterilized bottled medium, 0.2 ml of sterile sodium thiosulphate solution, 0.05 ml of sterile biotin solution (2.0 mg in 100.0 ml H_2O) and 0.05 ml of filter sterilized *p*-aminobenzoic acid (10.0 mg in 100.0 ml H_2O) are injected. For *Rhodopseudomonas sulfoviridise*, 0.1% sodium malate and 0.05% sodium thiosulphate are added and the pH is adjusted to 6.8. After sterilization, neutralized sterile sodium sulphide solution is added to a final concentration of 0.03%.

For the control of anaerobiosis, 1–2 drops of resazurin solution (5 mg in 5 ml H_2O) are added to a few representative bottles before autoclaving. This turns the medium blue. During autoclaving, residual oxygen combines with the ingredients in the medium and the resazurin is reduced to a colourless or pink form (redox potential below -50 mV) without the addition of any other reducing agents.

Trace element solution SL-6:

$ZnSO_4.7H_2O$	0.1 g
$MnCl_2.4H_2O$	0.03 g
H_3BO_3	0.30 g
$CoCl_2.6H_2O$	0.20 g
$CuCl_2.2H_2O$	0.01 g
$NiCl_2.6H_2O$	0.02 g
$Na_2MoO_4.2H_2O$	0.03 g
Distilled water	1000 ml

Neutralized sulphide solution:

Distilled water	100.0 ml
$Na_2S.9H_2O$	1.5 g

The sulphide solution is prepared in a 250 ml screw-capped bottle with a butyl rubber septum and a magnetic stirrer. The solution is bubbled with N_2 gas, closed and autoclaved for 15 min at 121°C. After cooling to room temperature the pH is adjusted to about 7.3 by adding sterile 2 M H_2SO_4 drop-wise with a syringe, without opening the bottle. Appearance of a yellow colour indicates the drop of pH to about 8. The solution should be stirred continuously to avoid precipitation of elemental sulphur. The final solution should be clear and is yellow in colour.

2. Modified Pfennig's medium for purple sulphur bacteria

Solution A:

$CaCl_2\ 2H_2O$	0.5 g
Yeast extract	0.5 g
Distilled water	925.0 ml

Solution A (46 ml) is distributed into 100 ml screw-capped bottles (with metal caps and autoclavable rubber seals). Each bottle is bubbled with N_2 gas or, better, a mixture of 95% N_2 and 5% CO_2, closed immediately and the cap screwed tightly. The bottles are autoclaved at 121°C for 15 min. For marine or estuarine isolates appropriate amounts of NaCl are added to this solution.

Solution B:

$Na_2S.\ 9H_2O$	2.0 g
Distilled water	135.0 ml

This solution is autoclaved in a screw-capped bottle under an atmosphere of N_2.

Solution C:

$NaHCO_3$	3.0 g
Distilled water	100.0 ml

This solution is bubbled with CO_2 and, after saturation, is filter sterilized under CO_2 pressure into sterile, gas-tight, 100 ml screw-cap bottles.

Solution D:

Ammonium acetate	0.5 g
Ammonium chloride	0.7 g
Sodium pyruvate	0.5 g
Glucose	0.5 g
$MgSO_4.7H_2O$	1.0 g
KH_2PO_4	0.7 g
KCl	0.7 g
Vitamin B_{12} solution (0.001% in H_2O)	5.0 ml
Trace element solution (SL 12 B) (see p. 87)	2.0 ml
Distilled water	50.0 ml

This solution is filter-sterilized.

Solution E:

Resazurin solution (0.005 g in 5.0 ml H_2O)	1.0 ml
Distilled water	800.0 ml

This is autoclaved for 15 min at 121°C in a 2 litre Erlenmeyer flask with an outlet near the bottom, to which a silicon rubber tube with a pinch cock and a bell is attached for aseptic dispensing of the medium into bottles. The solution E is placed in an ice bath and to it solutions C and D are added. It is bubbled with CO_2 for about 15 min. About 50 ml of this is distributed aseptically to each 100 ml screw-cap bottle containing 46 ml of solution A and the bottles are immediately closed tightly. For purple sulphur bacteria about 4 ml of solution B is injected into each bottle and the pH adjusted to about 7.2 with sterile HCl or Na_2CO_3 solution (2 mol l^{-1} each). The medium turns pink and later colourless, due to the reduction of the indicator resazurin.

A small air bubble is left in each bottle to meet possible pressure changes. The tightly sealed, screw-cap bottles can be stored for several months in the dark. During the first 24 h, the iron of the medium precipitates in the form of black flocks. No other sediment should arise in the otherwise clear medium. Before inoculation, it is recommended that the medium is allowed to stand for a few hours for equilibration and the detection of leakages. The medium will turn pink if the redox potential becomes higher than -100 mV.

Cultures are incubated in the light using a tungsten lamp. Feeding is done periodically with a neutralized solution of sodium sulphide (see medium for *Rhodospirillaceae*) to replenish sulphide, and with other supplement solutions (Malik, 1983, 1984b).

Trace element solution SL−12B:

Distilled water	1000.0 ml
Ethylenediamine tetra-acetate (Na$_2$-EDTA)	3.0 g
FeSO$_4$.7H$_2$O	1.1 g
CoCl$_2$.6H$_2$O	190.0 mg
MnCl$_2$.2H$_2$O	50.0 mg
ZnCl$_2$	42.0 mg
NiCl$_2$.6H$_2$O	24.0 mg
Na$_2$MoO$_4$.2H$_2$O	18.0 mg
H$_3$BO$_3$	300.0 mg
CuCl$_2$.2H$_2$O	2.0 mg

The pH is adjusted to 6.0.

3. Modified Pfennig's medium for green sulphur bacteria

The medium is prepared as for purple sulphur bacteria. In solution D, 2.0 ml of trace element solution SL-10B (see below) is added instead of SL-12B. After filling all solutions in 100 ml screw-cap bottles, 5 ml of solution B is injected into each bottle and the pH is adjusted to 6.8 with sterile HCl or Na$_2$CO$_3$ solution (2 mol l^{-1} each). All other steps are the same as for purple sulphur bacteria.

Trace element solution SL-10B:

Distilled water	1000.0 ml
HCl (25%)	7.7 ml
FeSO$_4$.7H$_2$O	1.5 g
ZnCl$_2$	70.0 mg
MnCl$_2$.4H$_2$O	100.0 mg
H$_3$BO$_3$	300.0 mg
CoCl$_2$.6H$_2$O	190.0 mg
CuCl$_2$.2H$_2$O	2.0 mg
NiCl$_2$.6H$_2$O	24.0 mg
Na$_2$MoO$_4$.2H$_2$O	36.0 mg

4. Modified growth medium for Chloroflexus

Yeast extract	1.0	g
Glycyl-glycine	1.0	g
Na$_2$HPO$_4$.2H$_2$O	0.1	g
MgSO$_4$.7H$_2$O	0.1	g
KNO$_3$	0.1	g
NaNO$_3$	0.5	g
NaCl	0.1	g
CaCl$_2$.2H$_2$O	0.05	g
Ferric citrate solution (0.1 g in 100 ml H$_2$O)	5.0	ml
Trace element solution SL-6 (see medium for Rhodospirillaceae)	1.0	ml
Distilled water	1050.0	ml

The pH of this modified (Castenholz and Pierson, 1981) medium is adjusted to 8.2. The medium is boiled under a stream of N$_2$ gas for a few minutes and 90 ml is distributed into 100 ml screw-capped bottles. Each bottle is bubbled with N_2 and closed immediately with a rubber septum and screw tightened. The bottles are autoclaved at 121°C for 15 min. After autoclaving, 1.0 ml of neutralized sulphide solution (0.015% end concentration, see medium for Rhodospirillaceae) is injected into each bottle. This medium can be stored for several months. Incubation is carried out at 50°C at a light intensity of 300–500 lx. To obtain a heavy cell suspension, cultures are supplemented periodically with sterile yeast extract solution (0.1% end concentration); for more details see Malik (1983, 1984b).

III. MAINTENANCE BY SUBCULTURING

Most phototrophic bacteria can be preserved as actively growing cultures if maintained at reduced temperatures and in dim light, using reduced media of lower redox potentials. In practice, actively growing cultures in gas-tight bottles are kept at 4–15°C in light refrigerators (at about 10 lx) with 12 h light and 12 h dark periods. This treatment allows better survival over longer periods.

Purple nonsulphur bacteria and *Heliobacterium* strains can be maintained longer as agar stab cultures in mineral media, or in media of low organic nutrient concentrations. During maintenance of green and purple sulphur bacteria, a continuous supply of sulphide is needed for growth. At early stages of growth, increased amounts of sulphide cannot be included into the medium, due to toxicity problems. However, as soon as the sulphide in the medium is exhausted it must be replenished immediately as, without a photosynthetic electron donor cultures will stop growing and are subsequently damaged by further illumination.

Freshly grown cultures of Chromatiaceae are supplemented with a neutralized sulphide solution to a final concentration of 1.5 mM and kept in the light for about 2–3 h until the cells have formed intracellular sulphur globules. For periodic feeding, the cultures are put back under dim light at room temperature and are supplemented with a neutralized sulphide solution, and returned to the refrigerator after formation of sulphur globules.

The green sulphur bacteria are incubated until they have used up the transiently formed elemental sulphur and are then stored in a refrigerator. After 2–3 months of storage they are supplemented again with a neutralized sulphide solution, incubated in dim light until sulphide and sulphur are consumed, and placed again in a refrigerator.

The liquid cultures are transferred to fresh culture media after 5–6 months of storage; however, cultures in or on agar media can be stored for longer periods.

The genus *Chloroflexus* resembles *Chlorobium* in cell structure and chlorophyll pigments, but is similar to the purple non-sulphur bacteria in being both photoheterotrophic and facultatively chemoheterotrophic, in addition to having the ability to grow photoautotrophically. Anaerobic conditions and the presence of sulphide favours growth and prolonged viability during maintenance as active living cultures. Thickly grown (but not old) liquid cultures can be maintained under anaerobic conditions for several weeks at 37–45°C in dim light (100–200 lx) if they are regularly supplemented with yeast extract. Photoheterotrophically grown shake or agar stab cultures may also be maintained under the above conditions.

The presence of activated charcoal in the reactivation media results in higher survival levels and leads to more stable cultures that can be maintained longer as active living cultures. The use of activated charcoal in the growth media is recommended as it is believed to absorb various harmful radicals and work as a neutral and harmless reducing agent, especially for anaerobic cultures that are sensitive to other toxic reducing agents such as sulphide and cysteine.

IV. MAINTENANCE BY FREEZING

Cryopreservation of microorganisms in liquid nitrogen at −196°C is a very reliable method and is generally considered superior to other preservation methods. Bacteria preserved in liquid nitrogen normally show high survival rates and good strain stability during long-term storage.

In liquid nitrogen storage of microorganisms, polypropylene cryotubes, glass vials, glass capillaries and polypropylene straws have been successfully used. In the method described here, mini-screw-cap glass ampoules have been used (Malik, 1984b) as they are suitable for the preservation of bacteria requiring anaerobic conditions. The method is simple, effective and economical with respect to storage space and costs. No continuous stream of N_2 gas, anaerobic chambers or glove boxes are required. Almost all existing species of phototrophic bacteria at the DSM have been successfully preserved with this method, and it provides a suitable model for cryogenic storage of many fastidious and delicate bacteria. The details of this method are described below (Fig. 1).

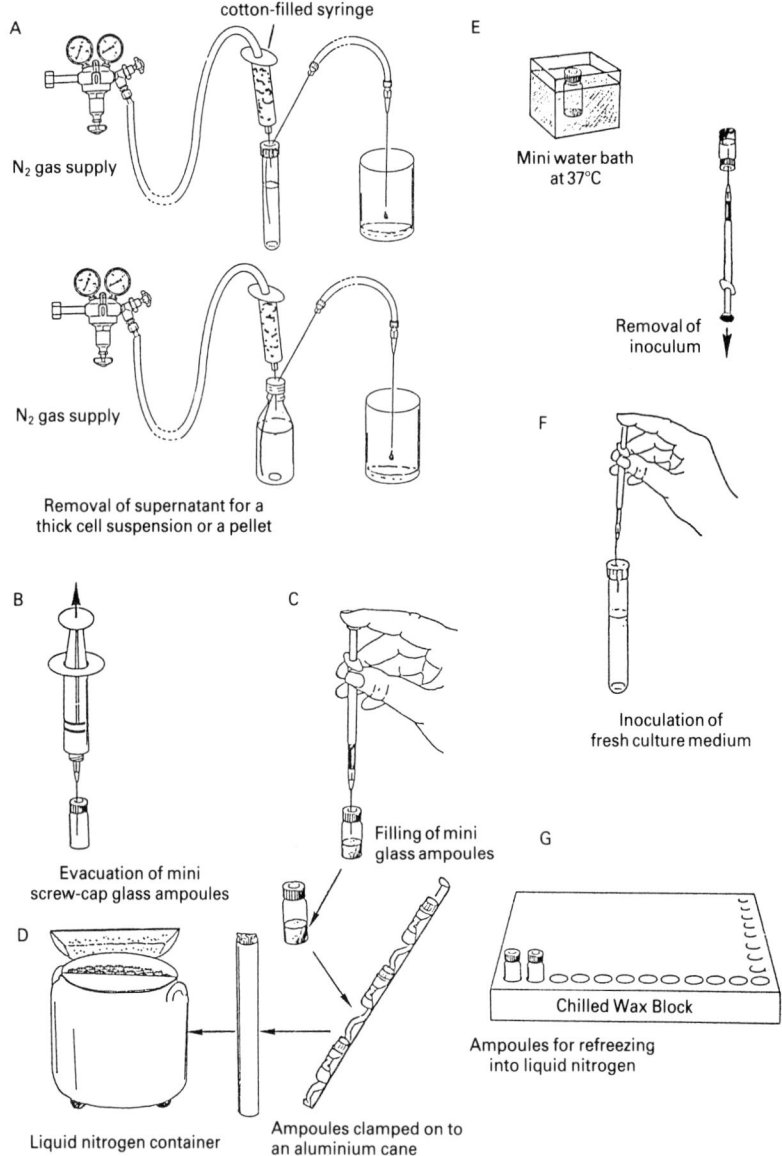

Fig. 1 Cryopreservation under aerobic conditions.

1. Equipment   Screw-cap glass ampoules (10 × 30 mm) of 2 ml capacity (available from Varian GmbH, Darmstadt, FRG) are used. These are generally used with autosamplers in gas chromatography and are provided with oxygen-impermeable butyl rubber septa, with holes for the injection of the samples, and black plastic screw-caps.

The ampoules are washed, rinsed with distilled water, tightly closed and autoclaved. Before use, they are labelled with the numbers of the strains to be preserved.

Liquid nitrogen storage tanks with cannisters, racks, canes, (supplied by Union Carbide; Air Liquide, France; Messer Griessheim, FRG) and related safety equipment such as glasses and gloves are required.

Anaerobic facilities for the preparation of reduced media are necessary. Hungate tubes with septa (Bellco Glass Inc., 2047–16125); butyl rubber overflow tubes (about 5 mm in diameter) with Luer Lock adapters at both ends and long syringe needles (10–15 cm in length); sterile gas tight hypodermic Luer Lock syringes.

Cryoprotective agents glycerol and dimethylsulphoxide (DMSO), reagent grade, may be used. Glycerol (20% w/v in H_2O) may be sterilized by autoclaving for 15 min and is stored in screw-cap bottles in the dark. DMSO (20% v/v in H_2O) is sterilized by filtration (using a Teflon syringe filter) or can be autoclaved undiluted at 114°C for 10 min.

2. Preparation of cell suspension for freezing

Members of the family Rhodospirillaceae are grown in appropriate media under anaerobic conditions and are harvested preferably in the active phase of growth. From agar slants, the growth is removed with a loop and gently suspended in sterile glycerol (10% w/v) or DMSO (5% v/v) to obtain a heavy cell suspension. For brown Rhodospirillaceae cultures, this is done under a stream of sterile N_2 gas. Thick suspensions (10^8–10^{10} cells ml^{-1}) of liquid cultures are mixed in equal quantities with the double concentrated cryoprotective agents.

The anaerobically grown liquid cultures are centrifuged for 30 min at 4000 g in the screw-cap bottles in which they have been grown. The supernatant is removed under a stream of N_2 gas using an overflow butyl rubber tube of about 5 mm diameter with Luer Lock adapters at both ends and fitted with long syringe needles of 10–15 cm length (see Fig. 1A). To obtain sterile nitrogen gas, a sterile, cotton-filled syringe is attached to a conduit connected to the N_2 gas (99.99%) cylinder. After centrifugation the sterile, long needle (attached to the overflow butyl rubber tube) is injected into the bottle and, simultaneously, the cotton-filled syringe with a small hypodermic needle (0·80 × 38 mm, already attached to the N_2 gas supply) is also injected into the bottle. Slight over-pressure due to the supplied sterile N_2 gas results in a smooth overflow of the supernatant through the butyl rubber tube to the nearby vessel. The long-needle is slowly lowered into the supernatant to remove the desired amount or to leave a pellet.

During this process the remaining cell suspension or the pellet remains anaerobic and the reduced state of resazurin (which is present in the medium as an anaerobic indicator) is maintained. The pellet is resuspended carefully in ice-cold sterile DMSO solution (5% v/v in H_2O). In the case of halophilic strains or cells which do not form a pellet, the thick bacterial suspension left in the growth medium is mixed in the ratio 3:1 with ice-cold sterile DMSO (20% v/v in H_2O). For extremely halophilic strains, optimum salt concentrations should be maintained after mixing the cell suspension with the DSMO. The cells are allowed to equilibrate with the cryoprotectant (15 min for DMSO, 30 min for glycerol) in an ice bath.

3. Filling ampoules and freezing

While equilibrating, an aliquot of 1.0–1.5 ml of cell suspension is dispensed into each glass ampoule. Using a sterile gas-tight, 5–10 ml syringe, the ampoules are evacuated for anaerobiosis and to facilitate filling (Fig. 1B). About 1 ml of thick cell suspension (equilibrated with the DMSO) is withdrawn with a 1 ml sterile oxygen-free syringe (already flushed with N_2 gas) and injected into each ampoule (Fig. 1C). The glass ampoules are immediately clamped onto labelled aluminium canes and are either placed at $-30°C$ for about 1 h or for a few minutes in the gas phase of liquid nitrogen to achieve a cooling rate of about 1°C per minute. The canes are then placed in cannisters, or the ampoules transferred to racks or drawers and frozen by direct immersion in liquid nitrogen or gas phase of liquid N_2 (Fig. 1D).

4. Revival of cultures

The frozen ampoules are removed from the liquid nitrogen refrigerator. For transportation, they are placed in a self-made wax block rack which has been cooled to $-30°C$. They are partially thawed by immediately immersing up to the neck of the ampoule in a mini water bath at 37°C (Fig. 1E) for a few seconds. The septum of the glass ampoule is flame-sterilized after addition of a drop of alcohol, and with a 1 ml oxygen-free syringe a small volume (about 0.05 ml) of inoculum is withdrawn and injected into 5–10 ml of liquid growth medium (Fig. 1F). The rest of the partially thawed cell suspension is immediately frozen again (by replacing the ampoules in the chilled wax block and transporting back to the liquid nitrogen container — see Fig. 1G) in liquid nitrogen for later use. In this way one ampoule can be used for several repeated retrievals or inoculations. The contents of the ampoule are immediately transferred to fresh growth medium to dilute the cryoprotectant 100–200 times to a non-inhibitory concentration; DMSO is often lethal at higher temperatures.

5. Estimation of viability counts

For Rhodospirillaceae, 0.5 ml of inoculum is transferred to 4.5 ml of liquid growth medium and serial decimal dilutions are prepared. Plating and counting are done using standard methods

(Malik, 1983, 1984b). For the estimation of viable cell counts in anaerobic phototrophic bacteria (green and purple sulphur bacteria), 0.5 ml of inoculum is transferred from the unfrozen (for cell counts before freezing) and from the thawed cell suspension (for cell counts after freezing) into pre-reduced 4.5 ml medium in screw-cap tubes (Hungate tubes with septa, Bellco Glass Inc., 2047–16125) and six to eight serial decimal dilutions are prepared, using oxygen-free syringes, and incubated under appropriate conditions.

Agar roll tubes (Hungate, 1969) can be prepared for viable colony-count determinations if such facilities are available. Colony counts on agar plates can be performed in an anaerobic glove box or in anaerobic jars. Single plates can be incubated in anaerobic Bio-bags (Type A, Marion Scientific Corp.) or in Anaerocult bags (E. Merck). In the case of viable colony counts in agar roll tubes or on plates, the number of colonies are counted from each dilution and the average colony forming units per sample are calculated. For cultures which are difficult to grow in or on agar, only liquid dilutions series are made. In this case, the number of cells is determined using the most probable number method (MPN). The percentage survival is calculated by comparing viable cell counts before and after freezing.

V. MAINTENANCE BY FREEZE-DRYING

Freeze-drying is a convenient method for the preservation and long-term storage of a wide variety of microorganisms (Malik, 1976; Lapage et al. 1970), see chapter 4.1. However, poor survival results when anaerobically grown cultures of phototrophic bacteria are preserved by freeze-drying under aerobic conditions. Although several species of Rhodospirillaceae are able to grow aerobically in the dark under heterotrophic conditions, strict anaerobic conditions are usually needed for their photoautotrophic growth and maintenance. During a systematic investigation it has been observed that such heterotrophically grown cultures are not damaged by air-oxygen and showed no loss in photoautotrophy when they were grown again in light under anaerobic conditions. Based on this observation, heterotrophically grown cultures of several species of Rhodospirillaceae can successfully be preserved by freeze-drying under aerobic conditions (Malik, 1988b).

Freeze-drying of anaerobes is often difficult (Staab and Ely, 1987). Some strains of anoxygenic phototrophic bacteria are very sensitive to oxygen and can only grow as photoautotrophs or photoheterotrophs under anaerobic conditions in the light. In such cultures the photopigment content (carotenoid, bacteriochlorophyll or porphyrin) is high and is responsible for sensitivity to light and oxygen (Malik, 1988b). Thus, failure to preserve such cultures by conventional aerobic methods is mainly due to their sensitivity to oxygen in the presence of light, presumably resulting in membrane damage due to photodynamic effects (Malik, 1984a).

Porphyrine solutions have been added to cell suspensions to absorb light during the process prior to freeze-drying and have been found to offer good protection against photooxidation during lyophilization (Malik, 1978). However, porphyrine solutions are difficult to prepare and a prolonged exposure to light causes them to generate single oxygen molecules, which are toxic. During a systematic investigation it was observed that activated charcoal protected phototrophically grown cultures against photooxidation by absorbing light and oxygen before freeze-drying. As a result, activated charcoal has been used instead of porphyrins, and successful freeze-drying of various phototrophs and other anaerobes has been achieved without the use of a glove box or an anaerobic chamber (Malik, 1990a). However, for freeze-drying of very sensitive anaerobes, standard anaerobic conditions are still required prior to freeze-drying.

The freeze-drying methods described here are based on a double-vial system in which the ampoules containing the freeze-dried material are sealed under vacuum in a soft glass tube.

A. Freeze-drying under aerobic conditions

Currently, 10 species of Rhodospirillaceae are able to grow heterotrophically in the dark on organic media. Such cultures can be lyophilized successfully using raffinose (5% w/v), with skim milk (20% w/v) as a protective agent (Malik, 1988b). All cultures remain viable and show a 10–100% survival after lyophilization. During 2–3 years of storage at 9°C no further loss in viability has been observed nor any loss of photoautotrophy, diazotrophy or other desirable characteristics such as hydrogen production or pigmentation. The details of this method are described below. It is not suited to the anoxygenic Rhodospirillaceae which are unable to grow aerobically in the dark.

1. Growth of cultures

Strains of *Rhodobacter capsulatus*, *Rhodobacter sphaeroides*, *Rhodocyclus gelatinosus*, *Rhodopseudomonas roseus* and *Rhodopseudomonas acidophila* are grown in nutrient agar slants and incubated at 25–30°C in the dark. The cultures of *Rhodobacter sulfidophilus*, *Rhodopseudomonas blastica*, *Rhodopseudomonas marina*, *Rhodopseudomonas palustris* and *Rhodospirillum rubrum* are grown in modified mineral medium (Malik, 1983), supplemented with an appropriate carbon source or in an organic medium (H3P) developed for the heterotrophic growth of a broad spectrum of aerobic bacteria (Malik, 1988b).

2. The freeze-drying procedure (Malik, 1988a)

Ampoules of freeze-dried skimmed milk containing a protective agent are prepared as described below.

The ampoules are labelled and filled with 0·5 ml of 20% (w/v) skimmed milk (Bacto, Difco 0032) containing 5% *meso*-inositol or raffinose. They are loosely plugged with non-absorbent cotton wool and sterilized at 115°C for 13 min, frozen at about −40°C for a few hours and are then freeze-dried in bulk for 12–24 h using a standard freeze-drying technique. A thick cell suspension (at least 10^8 cells ml^{-1}) is prepared in a filter-sterilized aqueous solution of raffinose (5% w/v) and kept in an ice bath for about 15 min for equilibration. One drop (0.05 ml for precise viability count determinations) of this suspension is added to each pre-chilled freeze-dried skimmed milk ampoule and frozen at approximately 1–2°C per minute to about −30°C.

For primary freeze-drying, the refrigerator of the freeze-drying machine (Edwards Freeze-dryer Modulyo) is switched on, and the condenser drain valve is closed. When the condenser temperature falls to below −50°C, the frozen ampoules are transferred quickly to the freeze-drying chamber, vacuum is applied and the air admittance valve is slowly closed. Primary freeze-drying is continued at 0·05 Torr for 12–16 h to achieve maximum desiccation. Ampoules that are to be sealed under vacuum are transferred to soft glass tubes (130 × 15 mm) containing silica gel and cotton plugs. The outer vials are then constricted (using an Edward's Ampoule Constrictor) and are attached to a manifold mounted on a freeze-drying machine. The vacuum is switched on and secondary drying is conducted for 3–4 h (at 0.001 Torr). After this time the constricted outer vials (containing inner vials) are sealed under vacuum using a Flaminaire blow torch.

3. Estimation of viability and stability testing

Survival levels are checked before freeze-drying, immediately after freeze-drying and after 2–3 years of storage. For the estimation of viability counts exact volumes (0.05 ml) of cell suspension are freeze-dried and used to prepare serial dilutions in appropriate liquid media. From each serial dilution 0.1 ml is transferred to plates of appropriate heterotrophic growth agar and incubated in the dark under aerobic conditions. From each serial dilution, 0.1 ml volumes are also plated in parallel on mineral agar medium and viability counts are performed under anaerobic photoautotrophic conditions as described earlier. The revived cultures are observed for changes in pigmentation or colony morphology.

B. Freeze-drying under anaerobic conditions

For anaerobically grown cultures of anoxygenic phototrophic bacteria it is essential to maintain anaerobic conditions and to avoid the exposure of phototrophic cultures to light during the freeze-drying process. Reducing agents, such as sulphide, cysteine and amorphous ferrous sulphide, have been used, but the former two proved toxic and lethal. During freeze-drying, where evacuation is involved, toxic vapours can increase to lethal concentrations. Although neutral amorphous ferrous sulphide has proved a satisfactory reductant before freeze-drying, it is tedious to prepare and its presence in the reactivation medium for phototrophic bacteria is undesirable.

It has been observed that activated charcoal protected photo-trophically grown cultures against photooxidation by absorbing light and oxygen before freeze-drying. Using activated charcoal and appropriate protective agents, several anaerobic purple non sulphur bacteria, such as Chromatiaceae and Ectothiorhodo-spiraceae, which cannot grow aerobically in the dark, and green sulphur bacteria were successfully preserved (Malik, 1990a). With the exception of *Rhodobacter purpureus* and a few strains of *Rhodospirillum molischiaum* and *Rhodospirillum photometricum*, almost all Rhodospirillaceae proved viable after lyophilization with a fairly good survival level. The survival of some strict anaerobic phototrophic bacteria after freeze-drying and storage showed a maximum loss of 1–2 $\log_{10}$ counts in many cases. This represents a relatively good survival recovery as similar losses may occur even during the cryopreservation of such fragile bacteria. However, *Heliobacterium*, and some sensitive strains of *Chromatiaceae*, *Chloroflexus* and *Chlorobiaceae* cannot be freeze-dried using this method. Details of the method are described below.

1. Growth of cultures

Strains of anoxygenic phototrophic bacteria are grown under anaerobic phototrophic conditions in the light and are fed periodically to obtain heavy cell suspensions.

2. Preparation of reduced protective agents

A solution either of *meso*-inositol (10% w/v) or raffinose (10% w/v) is prepared together with neutral activated charcoal (10% w/v) in distilled water. The activated charcoal used is of medicinal grade (obtained from Caelo, FRG), but any other bacteriological activated charcoal of comparable quality can also be used. The solution is boiled, bubbled with N_2 gas and closed tightly and auto-claved at 115°C for 13 min. This can be stored for several weeks.

3. Preparation of cell suspension

For harvesting, well-grown cultures are centrifuged under aseptic anaerobic conditions for 30 min at 4000 *g* in the screw-cap bottles in which the cultures have been grown (Malik, 1984b).

The supernatant is removed anaerobically as described under liquid nitrogen preservation (Section IV.2 above). To the thick cell suspension (10^8 to 10^{10} cells ml^{-1}) thus obtained in the closed nitrogen-filled screw-cap bottle, equal amounts of the sterile reduced protective agent mixture (10% w/v *meso*-inositol containing 10% w/v activated charcoal) is injected and mixed. The suspension is kept in an ice bath for 15 min for equilibration.

4. The freeze-drying procedure

For a double vial preparation, glass ampoules are labelled and filled with 0.5 ml of 20% w/v skimmed milk (Bacto, Difco 0032) containing 5% w/v *meso*-inositol or 5% w/v raffinose and 10% w/v activated charcoal. The ampoules are loosely plugged with non-absorbent cotton wool and sterilized at 115°C for 13 min. They are frozen at -40°C for a few hours and then freeze-dried in bulk for 12–24 h using a standard freeze-drying technique (Lapage *et al.*, 1970; Malik, 1988a). The prepared ampoules are labelled and are frozen at -30°C for 1–2 h. To each ampoule, two or three drops (about 30 µl) of reduced (equilibrated) cell suspension are added using a 1 ml oxygen-free syringe (with a 0.80×38 mm hypodermic needle). The inoculated ampoules are then frozen at approximately 1–2°C per minute to -30°C.

For strains very sensitive to oxygen, the prepared freeze-dried skimmed milk ampoules are first placed for 2–3 h in anaerobic Bio-bags (Type A, Marion Scientific Corporation) or in Anaerocult P bags (E. Merck). These are carefully taken out one by one for inoculation and, after the addition of the reduced suspension, the ampoules are immediately transferred to fresh anaerobic bags, which are sealed and kept for about 1 h at room temperature for equilibration and removal of oxygen. Thereafter, these cultures are frozen (still in the bags) at -30°C. The frozen ampoules are removed from the bags and transferred quickly to the freeze-drying chamber (Edwards Freeze-dryer Modulyo) for primary freeze-drying. Primary freeze-drying is continued as described previously (Malik, 1988a,b) at 0.05 Torr for 12–16 h to achieve optimum desiccation. Upon completion, the vacuum is replaced with N$_2$ gas (in the case of strict anaerobes) and the ampoules are constricted, subjected to secondary drying for 3–4 h (at 0.001 Torr), and are sealed under vacuum.

5. Revival of cultures from freeze-dried ampoules

The contents of the ampoule are rehydrated in pre-reduced sterile liquid growth medium and injected into 40–50 ml of growth medium. Freshly inoculated phototrophic cultures are placed for a few hours in the dark at an appropriate incubation temperature and later transferred to the light. A few cultures may exhibit a prolonged lag period and must be incubated for relatively longer periods. Normal growth usually appears after a second transfer into fresh medium. In a few cases growth is inhibited by the high concentration of the protective mixture used during lyophilization.

6. Estimation of viability and stability   For estimation of viability levels, special ampoules are inoculated with an exact volume (50 µl) of cell suspension. This is freeze-dried and, from the material recovered from such ampoules, serial dilutions are prepared in appropriate liquid media. The stability of cultures is checked by examining reactivated cultures for change in pigmentation and colony morphology. The viability counts of phototrophic bacteria are performed under anaerobic photoautotrophic conditions as described earlier (Section IV.5).

VI.   MAINTENANCE BY LIQUID-DRYING

Liquid-drying involves the vacuum-drying of samples from the liquid state without freezing. Several cultures of microorganisms that are sensitive to freezing or freeze-drying can successfully be preserved by liquid drying. At the DSM a simple and effective liquid-drying method has been developed using activated charcoal as a carrier material. In combination with other additives, activated charcoal serves as an ideal adsorption base and as a protectant against photo-oxidation. The method has successfully been applied for the preservation of a comprehensive collection of sensitive bacteria (Malik, 1990b). Several anoxygenic light and oxygen sensitive phototrophic bacteria such as Chromatiaceae, Ectothiorhodospiraceae, brown Rhodospirillaceae, green sulphur bacteria, *Chloroflexus* and *Heliobacterium* have been successfully dried in the presence of activated charcoal. Almost all tested strains retained viability and proved stable after L-drying and during storage at 9°C. The details of this method are described in Chapter 4.10.

VII.   GENERAL NOTES

Several factors can affect cell viability and stability during preservation. Protective agents are generally used during the preservation process as unprotected freezing, drying or freeze-drying may result in the complete desiccation of cells and thus damage to DNA. Good protective agents may prevent intracellular ice-crystal formation, and total desiccation as well as neutralizing harmful free radicals and electrolytic effects.

Freeze-drying is a widely used technique and there are many variations in technical details. Even good additives may prove less effective if freeze-drying is done under inadequate conditions. The method described here is based on slow freezing of cell suspensions (c. 1°C per minute) to about −30°C, prior to prolonged primary freeze-drying and short (2–3 h) secondary drying. It provides a suitable model for the freeze-drying of many delicate strains and is applicable to a broad range of microorganisms.

Successful cryopreservation can be achieved by the use of appropriate cryoprotective agents, maintaining a controlled rate of cooling and using an appropriate rewarming protocol. In practice, a relatively slow cooling rate can be easily obtained by keeping ampoules/vials in mechanical deep-freezers for 1–2 h or in the gas phase of the liquid nitrogen storage unit for some minutes before lowering the containers into liquid nitrogen. If cultures are stored submerged in liquid nitrogen, the liquid may seep into imperfectly closed or sealed containers. On removal from storage, nitrogen inside an ampoule may instantly change to the gaseous phase causing an explosion. For safety reasons it is recommended that cultures should be protected by a polythene sleeve (see Ch. 6A, Section VI) or that the straw method is used (see Ch. 6B). Alternatively, only if the stability of cultures is not affected by storage at higher temperatures, cultures may be stored in the gas phase of liquid nitrogen.

While preparing cells for cryopreservation several factors such as optimal growth conditions, physiological state of the cells (preferably from the late logarithmic to early stationary phase of growth) and a high cell density (10^6 to 10^8 cells per ml) should be considered as these can affect cell viability after cryopreservation. Cell suspensions should be allowed to equilibrate with the cryoprotective agent before freezing. For harvesting from liquid cultures, vigorous pipetting and high-speed centrifugation should be avoided and cells should be handled gently.

Safety precautions should be observed at all times. A face shield, laboratory coat and insulated gloves should be worn as protection against liquid nitrogen splashes and exploding ampoules. The level of liquid nitrogen in the containers should be checked, preferably on a daily basis, and maintained at a constant level, as any drop in liquid nitrogen level below a critical volume can result in damage due to the warming of the samples. The room in which the liquid nitrogen containers are kept should be well ventilated to prevent oxygen levels falling dangerously. The use of a warning oxygen meter should be considered.

During several years' experience it has been observed that when preserved (lyophilized or cryogenically stored) sensitive microorganism cultures are revived on agar media, the counts are usually lower than in liquid media. Similarly, agar media of high surface tension (such as nutrient agar) may lead to lower viable counts than mineral media of relatively lower surface tension.

During reactivation, the presence of activated charcoal in the suspending media results in higher survival, and the cultures are more stable and can be maintained longer as living cultures. Activated charcoal is not only effective against photooxidation but also creates a reduced environment. The presence of activated charcoal in the dehydration medium for anaerobes has also proved useful as an absorbent for various harmful radicals (sulphite, superoxide and other free radicals created by electron donors in reaction with oxygen). Due to its high adsorption capacity, charcoal stabilizes the extreme pH changes during growth and reduces the surface tension of dehydration media.

BACTERIA

4.8. MAINTENANCE OF METHANOGENIC BACTERIA

H. HIPPE

DSM — Deutsche Sammlung von Mikroorganismen
und Zellkulturen GmbH, Braunschweig, FRG

I. Introduction 101
II. Subculturing 103
   A. Method...................................... 104
III. Freezing 104
   A. Freezing in glass ampoules 105
   B. Freezing in glass capillary tubes 107
IV. Freeze-drying 110

I. INTRODUCTION

Methanogenic bacteria form a diverse group of procaryotic organisms physiologically united by the ability to produce methane as a result of their energy metabolism. They differ from the classical bacteria in such characteristics as their cell-wall polymers, membrane lipids, sensitivity to antibiotics, unique cofactors and sequence of 16S rRNA; they are known as archaebacteria (Balch et al., 1979; Hilpert et al., 1981; Kandler, 1982; Jones et al., 1987).

Methanogenic bacteria are obligate anaerobes; most of which are very sensitive to oxygen even when briefly exposed. Spores, or other resistant structures which could otherwise help survival under adverse conditions, are not known.

Currently, 57 species belonging to 18 genera have been validly described. With few exceptions, all methanogenic bacteria grow on an $H_2 + CO_2$ gas mixture; in addition many of them utilize formate and some grow on acetate, methylamines, methanol or a few other simple alcohols. The optimum temperature for growth of the various methanogens covers a remarkably broad range, from 20–25°C up to 100°C. The nutritional requirements are diverse; some strains are able to grow autotrophically in simple mineral media, whereas others require vitamins, cysteine,

MAINTENANCE OF MICROORGANISMS 2nd Edn
ISBN 012–410351–0

branched chain fatty acids, coenzyme M, acetate or unknown growth factors present in certain peptones or in rumen fluid. There are several marine species, some other species which require 7–10% or even 25% sodium chloride for optimal growth (as with extremely halophilic bacteria) and a few alcaliphilic species as well.

Three media have been described which fulfil the requirements of many of the existing methanogenic pure cultures (Balch et al., 1979). The preparation of media, as well as the successful cultivation of these bacteria, depends on the use of special equipment and techniques (Hungate, 1950, 1969; Macy et al., 1972; Miller and Wolin, 1974; Balch and Wolfe, 1976; Zeikus, 1977). Three types of culture vessels are used for growing methanogenic bacteria in small volumes: the screw-capped anaerobic culture tubes described by Hungate (Bellco Glass Inc., 2047–16125), the Balch serum tubes (Bellco Glass Inc., 2048–18150) and serum bottles. Both the latter are sealed with butyl rubber septa or stoppers held in place with crimped aluminium seals (Wheaton Scientific Div., 224183 and 224303). They are especially convenient for the growth of methanogens in a pressurized atmosphere of $H_2 + CO_2$ at 200–300 kPa which reduces the frequency with which the consumed gas mixture must be replaced (Balch and Wolfe, 1976; Balch et al., 1979). The screw-capped 'Hungate tube' can be pressurized to 100–200 kPa overpressure.

'Hungate tubes' containing 3–4 ml of agarized medium are also convenient for preparing dilution series in roll tubes, which are used for viability determinations by colony counts.

Agar plates can be used if a reliable anaerobic glove box is available. Plates are poured and inoculated inside the box, placed in a stainless steel anaerobic jar and removed from the box for incubation. Such anaerobic jars are equipped with a manometer and valves and can be pressurized with oxygen-free gas mixtures (Balch et al., 1979).

However, some methanogens appear to be unable to grow in or on the surface of agar media, and polysilicate plates have been developed for isolating these strains (Setter et al., 1981; Wildgruber et al., 1982). The ability of pure strains to grow on different kinds of agar has not been tested. Gelrite has been used successfully instead of agar, mainly for thermophilic strains (Belay et al., 1984).

Methods for maintaining cultures of the methanogenic bacteria may be grouped into: (i) subculturing, (ii) freezing, and (iii) freeze-drying.

II. SUBCULTURING

In general, the viability of most methanogenic bacteria can be prolonged when actively growing cultures are removed from the appropriate incubation temperature and kept at room temperature or refrigerated at 4°C. In this respect there is no major difference from non-archaebacterial anaerobes.

Cultures stored at reduced temperature remain viable for some weeks, provided no oxidation of the medium occurs. Species of the rod-shaped genera *Methanobacterium* and *Methanobrevibacter*, which are usually grown with an $H_2 + CO_2$ gas mixture, appear generally more resistant than species of other genera (Zeikus and Wolfe, 1972; Zeikus and Henning, 1975; Zehnder and Wuhrmann, 1977). There is no information available on survival times of formate-grown cultures.

In contrast to most other anaerobes, the methanogenic bacteria are grown either in media with relatively low organic nutrient content or in mineral media. Compared with rich media, simple media are not well protected against traces of oxygen, which may diffuse through the rubber septa or stoppers during storage of stock cultures. Pressurizing the culture vessels with the appropriate gas mixture before storing them at reduced temperature reduces the possibility of oxygen entering. Cultures of *Methanosarcina barkeri* grown on methanol or methylamines lose their activity after depletion of substrate, at a rate depending on the strain, and some need to be subcultured at intervals of 3–7 days. However, when grown with acetate or $H_2 + CO_2$ they often remain viable for several weeks. Methanococci tend to lyse rapidly after cessation of growth in liquid media at the optimal temperature. Cultures grown at 5–10°C below the optimum temperature and stored at room temperature may remain viable for up to 3 weeks (Whitman *et al.*, 1982; Whitman, 1989). *Methanococcus vannielii* requires weekly transfer when grown on formate (Jones and Stadtman, 1977). Again, *Methanothermus fervidus*, a highly thermophilic methanogen with an optimal growth temperature of 85°C, dies within a few hours after exhaustion of the $H_2 + CO_2$ gas atmosphere. The atmosphere must be renewed, therefore, or the culture transferred before the gas mixture is completely depleted (Stetter *et al.*, 1981). The following technique has been found useful for maintaining a number of methanogens grown in liquid culture with an $H_2 + CO_2$ gas mixture (Stetter *et al.*, 1981).

A. Method

Fresh medium is inoculated with 2–5% (v/v) inoculum from an active culture. After an appropriate incubation time (*c.* 6 h for the fast-growing strains; overnight for the slow-growing strains) at the optimum temperature, the culture vessel is repressurized with the same gas mixture to 200 kPa and stored at 4°C. Such cultures may be viable for several months. Samples from these cultures are either used to inoculate fresh media or, when a larger volume of an active culture is required, the whole of the stored culture is directly incubated. The applicability of this storage technique has not yet been proved for all types of methanogens.

III. FREEZING

As with other bacteria, freezing and storage at low temperatures is the most simple and effective way to preserve methanogenic bacteria. Several strains have been found viable after more than 5 years following freezing with 50% glycerol as cryoprotectant and storage at −70°C (W. E. Balch, pers. com.). With this method, equal volumes of mid-logarithmic cultures in 15 ml serum vials are mixed with sterile de-gassed and pre-reduced (0.05% cysteine plus 0.05% sodium sulphide) glycerol. After being pressurized to 200 kPa of nitrogen, the vials are placed in the freezer at −70°C.

Winter (1983) described a method to maintain methanogenic bacteria up to at least 20 months at −18°C with glycerol as cryoprotectant. The problem of oxygenation during storage through butyl rubber stoppers, which in the cold become very permeable to oxygen, could be prevented by enclosing the small rubber sealed vial containing the cell suspension in an outer, heat-sealed glass ampoule filled with oxygen-free gas. Freezing and storage of anaerobic agar cultures or biphasic agar–liquid cultures without glycerol at −76°C for up to one year has been described for strains of *Methanobrevibacter* and *Methanosphaera* (Miller, 1989).

The method of choice for the long-term preservation of cultures of methanogens is to freeze and store them in liquid nitrogen. Formate grown cultures of *Methanococcus vannielii* have been maintained by freezing in liquid nitrogen supplemented with sucrose to a 10 to 20% final concentration (Jones and Stadtman, 1977). Again, excellent results have been obtained at the Deutsche Sammlung von Mikroorganismen (DSM) with over 120 strains representing all existing species. The procedure used at the DSM is described below.

A. Freezing in glass ampoules

1. Preparation of cell suspension

Cultures are grown in heavy-walled, round-bottomed bottles (*c.* 70 ml volume) with necks that can be closed with a butyl rubber septum and a screw-cap as with Hungate anaerobe tubes (Bellco Glass Inc., 2047–16125). Such bottles can easily be made by a glass-blower. They fit in a normal laboratory centrifuge and are used both for growing the cells and for centrifugation.

Cells are cultured with an $H_2 + CO_2$ gas mixture in 20 ml of medium per bottle. Methanol or methylamines are used as growth substrates in 50 ml of medium per bottle for strains of *Methanosarcina*, *Methanococcoides*, *Methanohalophilus* and *Methanolobus*.

Cultures are grown to an optical density of about 0.300 at 600 nm, by which time about 4 ml of gas per millilitre of culture has been consumed. Cultures that use methyl compounds are harvested before active gas production ceases. The cultures are centrifuged directly in the unopened bottle, after which the bottle is opened, a gassing cannula inserted and the supernatant aseptically removed as completely as possible with 20 ml hypodermic syringe and a long needle of 2 mm diameter. To the cell pellet, 2 ml of suspending medium is added and cells are suspended by means of a Pasteur pipette, bent and heat-sealed at the tip. Depending on the number of ampoules that have to be prepared, cell suspensions from several cultures are collected in one bottle by transferring with a sterile Pasteur pipette.

2. Suspending medium and cryoprotectant

Fresh culture medium containing 10% (v/v) glycerol or 5% (v/v) dimethyl sulphoxide (DMSO) is used as a suspending medium. It is prepared just before use by adding 0.55 ml of sterile glycerol or 0.27 ml of sterile DMSO to 5 ml of sterile fresh medium in a culture tube. The cryoprotectant-containing medium is then reduced by adding appropriate amounts of cysteine hydrochloride and sodium sulphide from anaerobic stock solutions.

The cryoprotectant is sterilized separately by autoclaving in anaerobe test tubes under nitrogen atmosphere.

3. Filling ampoules

Clean, sterile, dry glass ampoules, such as the 2 ml Wheaton gold-band cryule (Wheaton Scientific Div., 651486) are filled with 0.1 or 0.2 ml of cell suspension using a calibrated Pasteur pipette connected to a pipette aid depositing the suspension at the bottom of the ampoule. Care is taken not to touch the inner side of the ampoule. The pipette is flushed with oxygen-free gas before aspirating the cell suspension. Similarly, the ampoule is flushed during filling by inserting a gassing cannula near the bottom. A mixture of $N_2 + CO_2$ is used instead of the H_2-containing gas mixture. After filling, the gassing cannula is moved near the top of the neck; then, while gassing is continued, the ampoule is heated in the middle of the long neck with a thin, hot flame from a gas burner, drawn out and sealed. Each ampoule prepared this way is immediately placed in an ice bath to prevent further warming of the contents from the heated upper part of the ampoule.

An amount of cell suspension equal to that used to fill the ampoules is transferred to 5 ml of fresh medium and used for the determination of pre-storage viability.

4. Freezing and storage

Methanogenic bacteria do not need to be frozen at a carefully controlled cooling rate if the suspensions are protected by glycerol or DMSO. Freezing, therefore, is simply achieved by placing the ampoules (dried outside after being removed from the ice bath) in the cold gas phase of the liquid nitrogen, using the storage system (aluminium canes or drawers) supplied with the refrigerator. This method provides a cooling rate of about 50°C per minute and gives good survival for routine work. For safety, the glass ampoules should be stored only in the vapour phase and not immersed in liquid nitrogen.

5. Recovery and determination of viability

For recovery, an ampoule is removed from the liquid nitrogen (using protective gloves) and quickly thawed by immediately placing it in warm water, at 30–37°C. The ampoule is agitated until thawing is complete. The ampoule is dried outside and scratched with a diamond pencil at the constriction. This area is wiped with ethanol and flamed. The ampoule is broken open and the suspension immediately aspirated in a 1 ml hypodermic syringe (flushed with oxygen-free gas) and injected through the rubber septum of an unopened culture tube containing 5 ml of fresh medium.

For determination of viability, decimal dilutions are prepared in a series of nine tubes using the syringe technique. A culture tube containing 3.6 ml of molten 1.5% agar medium kept at 45°C is inoculated from each dilution and agar roll tubes are prepared. Both the liquid dilution series and the roll tubes are incubated under appropriate conditions. Colonies grown in the roll tubes can easily be counted under a stereo microscope. A comparison of the counts before and after freezing is used to calculate the survival ratio. If an anaerobic glove box is available agar plates can be used instead of roll tubes.

Some methanogens, for example *Methanoplanus limicola* and *Methanothermus fervidus*, will not grow on the surface of, or in, agarized medium (Stetter *et al.*, 1981; Wildgruber *et al.*, 1982). Furthermore, the preparation of agar roll tubes may be laborious and time-consuming in routine work. In this case, only the liquid dilution series is made. After incubation, a rough estimation of the viability is made based on the last tube-yielding growth. With the slow-growing methanogens, the incubation period has to be extended to several weeks.

B. Freezing in glass capillary tubes

The use of glass capillary tubes instead of glass ampoules in the freeze-preservation of bacteria and protozoa has been described by Pautrizel and Carloz (1952), Cunningham et al. (1963), Walker (1966), and Jarvis et al. (1967). In combination with the 'Hungate technique', the glass capillary method has proved to be equally convenient for preserving strict anaerobes like the methanogenic bacteria. About 60 strains of non-methanogenic archaebacteria, including extreme thermophiles and extreme halophiles, are also currently maintained at the DSM using this technique.

Glass capillary tubes have several advantages compared with other miniaturized methods such as the plastic straws or glass-bead techniques (Nagel and Kunz, 1972; Feltham et al., 1978; Gilmour et al., 1978; Kirsop and Henry, 1984; Challen and Elliot 1986; Stalpers et al., 1987). They can be sealed hermetically and, because of their impermeability to gases, anaerobes are well protected against oxidation despite the large surface/volume ratio of the enclosed cell suspension. For recovery, a single capillary can be removed without thawing or contaminating the parallel samples. Again, glass capillaries can be stored in the vapour phase of liquid nitrogen as well as in the liquid phase, and are heavy enough not to rise to the surface when immersed. However, a tight seal is very important if capillaries are to be stored in the liquid phase. As with glass ampoules, improperly sealed glass capillaries take up nitrogen and will explode on removal from the cold. However, a perfect seal is easier to achieve with capillaries than glass ampoules.

Figure 1 shows the procedure for the glass capillary method as used for methanogens and other bacteria at the DSM for nearly 15 years.

1. Method   The cell suspension is prepared as described above. About 1 ml of the suspension is transferred into a small sterile vial which is kept anaerobic by gassing with the appropriate gas mixture. The vial is placed in an ice bath. One glass capillary (Difco Laboratories, 0658–33) is taken from the sterile stock by fitting it to the tip of a micropipetter (Rudolf Brand) and enough of the cell suspension is aspirated to fill one-third of the length of the capillary. The volume taken up is c. 0.025 ml. The suspension is aspirated until it is 1 cm from the free end of the capillary which is sealed in a fine, hot gas flame. The second seal is made

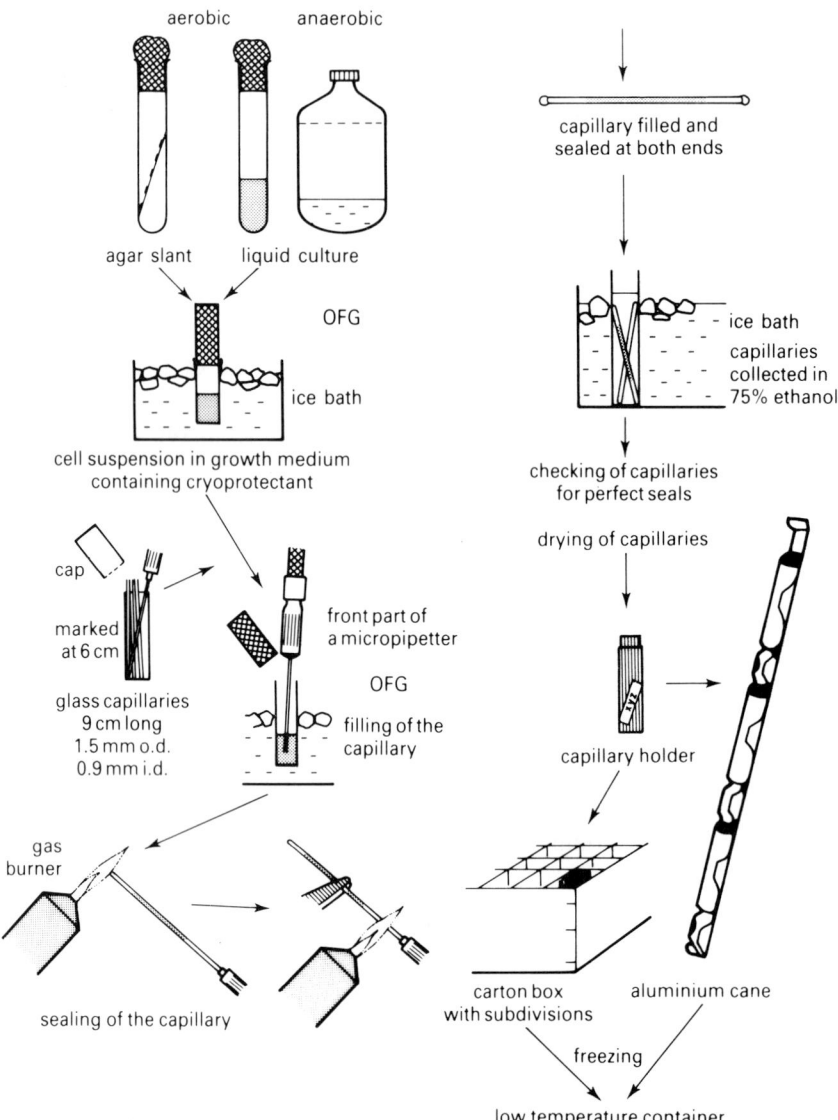

aerobic   anaerobic

agar slant   liquid culture

OFG

ice bath

cell suspension in growth medium
containing cryoprotectant

cap

marked
at 6 cm

glass capillaries
9 cm long
1.5 mm o.d.
0.9 mm i.d.

front part of
a micropipetter

OFG

filling of the
capillary

gas
burner

sealing of the capillary

capillary filled and
sealed at both ends

ice bath
capillaries
collected in
75% ethanol

checking of capillaries
for perfect seals

drying of capillaries

capillary holder

carton box   aluminium cane
with subdivisions

freezing

low temperature container

c. 2.5 cm away from the other end (a mark is made before
sterilizing the capillaries) by heating and drawing out.

As capillaries are prepared, they are placed in 75% ethanol
or a disinfectant solution. All capillaries are examined for proper
seals under a stereo microscope. Because moist capillaries will
freeze together when cooled, they are dried by placing them
between absorbent paper and gently pressing and rolling with
the flat of the hand. The capillaries are placed in a small vial
(quiver) that is labelled on the outside and left open. This is

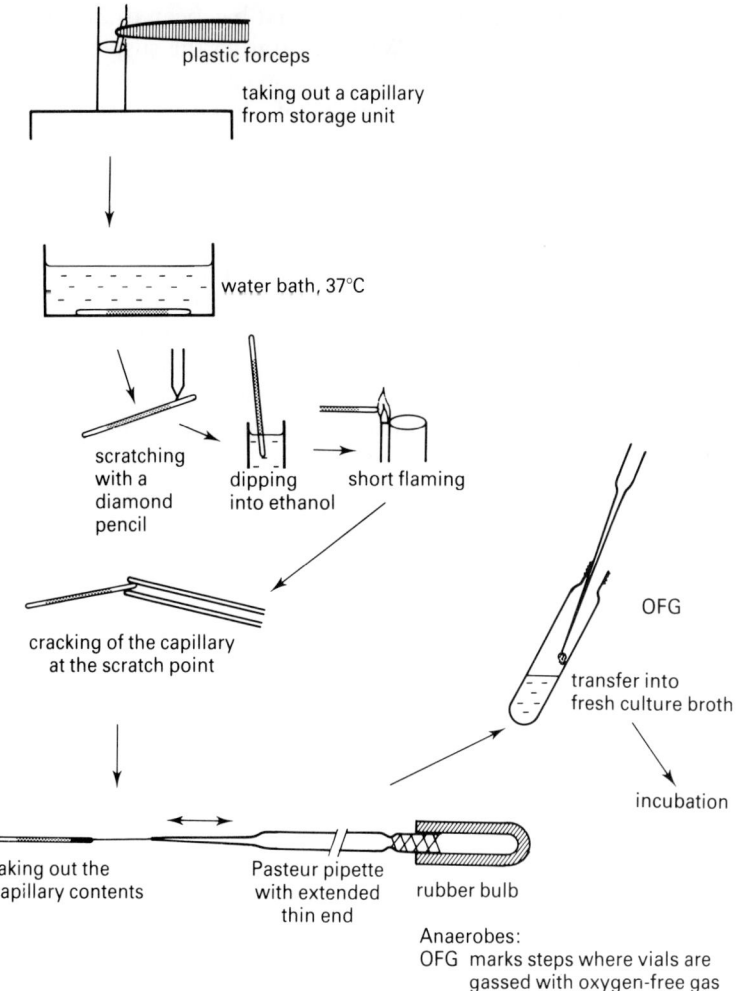

Fig. 1.   Left: Glass capillary tube method for low-temperature preservation of microorganisms — filling and sealing of capillary tubes. Right: Glass capillary tube method for low-temperature preservation of microorganisms — removal of tubes from freezing storage, opening, and recovery of cell suspension.

placed in the vapour phase of liquid nitrogen for freezing and storage.

Shock freezing by immersing capillaries directly in liquid nitrogen should be avoided because not all capillaries will withstand rapid cooling and some will break.

2. Recovery   A capillary is removed from the refrigerator and thawed rapidly in warm water. The thawing rate has been calculated to be over 1000°C per minute. The capillary is opened at one end as shown in Fig. 1 right. The small volume of cell suspension is aspirated with a sterile Pasteur pipette that has been drawn out very finely to a length of 4 cm. While aspirating the suspension, the pipette is slowly moved further into the capillary. A little experience is needed to avoid the uptake of air bubbles along with the cell suspension, which is then transferred to an opened culture tube containing fresh medium.

An alternative method for removing the cell suspension from a capillary is to use a 1 ml hypodermic syringe with a 0.50 or 0.65 mm needle. In this case the capillary is opened at both ends and the contents aspirated with a syringe previously flushed with oxygen-free gas. The needle is then inserted through the rubber closure of an unopened fresh culture tube. The tube is inverted and the suspension along with some medium is drawn into the syringe. Still in an inverted position, the contents of the syringe are ejected into the tube. This can be repeated once, but care should be taken not to expel gas bubbles into the tube if a plastic syringe is used.

IV. FREEZE-DRYING

Experience of freeze-drying methanogens is presently limited to only a few species. Studies performed on seven species of five genera have shown that members of the genera *Methanobacterium*, *Methanobrevibacter* and *Methanosarcina* in general survive the freeze-drying process better than members of *Methanococcus* and *Methanospirillum* (Hippe and Tilly, 1982). However, viable cultures could be recovered from them all after storage for 3–4 years.

Examples of viability levels of some methanogenic strains after freeze-drying and storage at two different temperatures for up to 42 months are given in Table I. It can be seen that, for some of the strains, optimal conditions of freeze-drying have not yet been achieved.

Suspending media, such as horse serum plus 7.5% glucose or 5% meso-inositol, 10% skimmed milk plus 10% sucrose, or 12% sucrose alone, which are widely used in freeze-drying other bacteria, are also suitable for methanogens. Inclusion of small amounts of amorphous ferrous sulphide (usually 1 mg ml^{-1}) with the suspending medium gave good protection of the cells against oxygen and allowed distribution of the small volumes of suspension in air, rather than in inert gas.

The following procedure based on the double vial method has been developed for methanogens.

Table I: Viability (cfu per sample) of some methanogenic bacteria after freeze-drying and storage.[a]

Species	DSM No.	Freeze-drying		Storage at		
					8°C	−70°C for
		Before	After	12 months	42 months	42 months
Methanobacterium formicicum[b]	1312	2.9×10^7	4.8×10^3	—	1.3×10^1	2.3×10^3
Methanobacterium thermoautotrophicum	1053	1.8×10^8	8.0×10^7	1.8×10^4	5.5×10^5	3.6×10^5
Methanobrevibacter arboriphilus	1125	8.0×10^{10}	5.9×10^{10}	1.3×10^{10}	1.0×10^{10}	1.1×10^{10}
Methanobrevibacter ruminantium	1093	9.4×10^9	3.2×10^5	6.6×10^5	5.9×10^5	4.8×10^4
Methanobrevibacter smithii	861	2.0×10^9	1.3×10^8	2.9×10^7	1.1×10^8	1.1×10^9
Methanococcus vannielii	1224	1.0×10^7	6.4×10^3	5.0×10^3	0	2.4×10^3
Methanosarcina barkeri	804	6.9×10^7	1.2×10^6	1.0×10^5	6.3×10^3	5.0×10^6
Methanospirillum hungatei	864	4.4×10^6	8.0×10^3	8.0×10^3	3.1×10^3	2.0×10^4

[a] The suspending medium contained horse serum and 7.5% glucose mixture with 3 mg ml^{-1} ferrous sulphide.
[b] Suspending medium for M. formicicum contained no ferrous sulphide.

1. Preparation of cell suspension	Cultures are grown and cell suspensions are prepared as described in Section III.A.1. The suspensions are kept under oxygen-free nitrogen gas until all vials have been filled.

2. Suspending medium

The horse serum–sugar mixture is prepared by dissolving 7.5 g of glucose or 5 g of meso-inositol in 100 ml of horse serum (Unipath Ltd, SR 35, natural clot). The solution is sterilized through a Seitz filter-pad and 5 ml volumes are dispensed under oxygen-free nitrogen in sterile anaerobe test tubes. Before use, 0.3 ml of concentrated suspension (*c.* 20 mg ml^{-1}) of amorphous ferrous sulphide in water is added.

The ferrous sulphide is prepared according to Brock and O'Dea (1977). Sterile stock suspensions, after autoclaving in sealed tubes under nitrogen gas atmosphere, can be stored at room temperature for several months.

3. Filling vials

Clean, cotton-wool plugged, labelled, sterile vials (11 × 45 mm; internal diameter 9 mm) are filled with 0.1 or 0.2 ml of cell suspension. Filling is carried out under air using a calibrated Pasteur pipette. An equal volume is used to inoculate a fresh culture tube for viability determination.

4. Pre-freezing

Suspensions are pre-frozen in a paraffin wax block before drying. The block (12 × 17 × 4 cm) is drilled with 24 holes 1.2 cm wide and 1.5 cm deep. The block, which is cooled overnight in a −20 or −70°C deep freezer, is filled with up to 24 vials and replaced in the freezer for a further 15 min. Still in the wax block, the vials are transferred to the vacuum dryer. The cooled block prevents thawing of the cell suspension during transport and subsequent drying.

5. Freeze-drying

The entire paraffin wax block containing the vials is placed under the plexiglass bell of a freeze-drying machine. Primary drying is carried out until the vacuum has dropped to 0.05 mbar and all free liquid has been removed. This takes about 4 h with 0.2 ml of cell suspension per vial. The drying process is then interrupted, the evacuated system is flooded with nitrogen gas and the vials are removed from the machine. The projecting parts of the cotton-wool plugs are cut off and each vial is placed in a second (outer) glass tube measuring 15 × 135 mm. The outer tubes contain a few pieces of self-indicating silica gel covered with a little cotton wool. The vials are covered with glass wool or stone wool slightly compressed to a layer 1–2 cm deep. The outer tubes are constricted above the inner vials and when cool are attached to the manifold of the freeze-drying machine for secondary drying overnight. At a vacuum of 0.05 mbar, the tubes are heat-sealed at the level of the constriction. The ampoules are stored in the dark at room temperature or preferably in a deep-freeze cabinet.

6. *Recovery and viability determination*

To open the double-vial ampoule, the pointed part of the outer tube is heated in a Bunsen flame. One or two drops of water are added to crack the hot glass and the tip of the tube is knocked off. The inner vial is taken out and immediately gassed with the appropriate gas mixture (H_2–CO_2 or N_2–CO_2). About 0.5 ml of fresh medium from an open gassed culture tube is added by means of a Pasteur pipette. The contents of the vial are agitated with the pipette to dissolve and resuspend the dried sample. After 2 min the contents are transferred to the culture tube. After sealing the inoculated tube, dilution series are prepared for viability determination as in Section III.A.5.

BACTERIA

4.9. MAINTENANCE OF LEPTOSPIRA

S. A. WAITKINS

Pathvet Services
Hereford, UK

I. Introduction 115
II. Subculturing 115
   A. Glassware 116
   B. Media 116
   C. Method 118
III. Liquid nitrogen 119
   A. Method 119
IV. Conclusions 120

I. INTRODUCTION

Leptospira are difficult to maintain for long periods; there is as yet no method which will guarantee a good recovery rate or preserve pathogenicity and antigenic pattern of the organisms. They may be kept alive, however, by time-consuming and tedious subculturing techniques or by storage in liquid nitrogen. All attempts at freeze-drying the organisms have been unsuccessful and earlier hopes of using this method have not been realized (Wolff, 1960; Otsuka and Manako, 1961a,b; Annear, 1962; Resseler *et al.*, 1966; Coghlan *et al.*, 1967).

II. SUBCULTURING

The frequency of subculturing depends on the medium used, the incubation temperature, and the dimensions of the container. In general, the narrower the tube and deeper the medium, the greater the likelihood of successful maintenance.

MAINTENANCE OF MICROORGANISMS 2nd Edn
ISBN 012–410351–0

A. Glassware

It is important that all glassware should be perfectly clean and free from soap or detergent residues. The cleaned glass should be soaked in Dulbecco phosphate buffer solution at pH 7.6 and rinsed in distilled water.

B. Media

Liquid or semi-liquid medium is used. Semi-liquid medium contains 0.5% agar (w/v) and allows longer intervals between subculturing, usually up to 12 weeks. Two media are recommended for maintenance and are outlined below.

1. Korthof's medium (Korthof, 1932)

(*a*) *Base A*

Neopeptone (Difco)	0.80 g
NaCl	1.40 g
$NaHCO_3$	0.02 g
$CaCl_2$	0.04 g
KH_2PO_4	0.24 g
$Na_2HPO_4.2H_2O$	0.88 g
Distilled water	1 litre

(*b*) *Base B (Inactivated rabbit serum).* Rabbit serum that is negative for leptospiral agglutinens should be used. Blood is collected from an ear vein and allowed to clot; the serum is then pipetted off, inactivated at 56°C for 30 min, and sterilized by Seitz filtration.

(*c*) *Preparation.* The ingredients of Base A should be steamed at 100°C for 20 min, cooled, and filtered through double thickness Whatman No. 1 paper (Scientific Supplies Co. Ltd.). The pH is adjusted to approximately 7.2. The whole volume is dispensed in 100 ml amounts and autoclaved at 115°C for 15 min. When cool, 8 ml of previously sterilized rabbit serum (Base B) is added aseptically to Base A and mixed thoroughly. The medium is distributed into either 150 × 13 mm glass tubes or sterile bijoux bottles with plastic caps; both should be tightly closed.

The dispensed medium should be incubated first at 37°C for 2 days and then at 22°C for a further 3 days to ensure the sterility of the product.

If semi-solid medium is used, 5 g of agar is added to Base A. Just before use two or three drops of fresh blood from a seronegative guinea pig are added to each container. When satisfactory growth has been achieved the tube is closed firmly with a rubber bung and stored either at 28°C or at room temperature (*c*. 22°C). Turner (1970) showed that using this medium leptospira survive up to 12 years.

2.
Ellinghausen
and
McCullough
(EM) medium
(Ellinghausen
and
McCullough,
1965a,b)

(*a*) *Stock solutions.* (The stock solutions are stored at −20°C until required.)

(i) Phosphate buffer, concentrated × 25

Na_2HPO_4	16.6 g
KH_2PO_4	2.172 g
Distilled water	1 litre

(ii) Salts, concentrated × 20

NaCl	38.50 g
NH_4Cl	0.35 g
$MgCl_2.6H_2O$	3.81 g
Distilled water	1 litre

(iii) Copper sulphate solution
  $CuSO_4.5H_2O$ 30 mg/100 ml distilled water

(iv) Zinc sulphate solution
  $ZnSO_4.7H_2O$ 80 mg/200 ml distilled water

(v) Ferrous sulphate solution
  $FeSO_4.7H_2O$ 500 mg/200 ml distilled water

(vi) Vitamin B_{12}
  (a) concentrate   10 mg/100 ml distilled water
  (b) working solution   10 ml concentrate + 90 ml
                distilled water

(vii) Vitamin B_1 200 mg/100 ml distilled water

(*b*) *Tween 80 (1% solution).* The Tween 80 stock bottle (BDH Chemicals Ltd) is placed in a water bath at 56°C together with a 1 litre flask containing 480 ml of distilled water. When both have equilibrated to 56°C, 4.8 ml Tween 80 is slowly pipetted into the warmed distilled water, taking care to rinse all traces of Tween 80 from the pipette. The 1% solution is shaken gently to mix.

(*c*) *Bovine albumin solution.* (This solution should be freshly prepared before use.)

	1 litre:
Pentax bovine albumin, Fraction V (Armour Pharmaceuticals)	50 g
Phosphate buffer concentrate (stock solution)	40 ml
Distilled water	960 ml

The distilled water is boiled and when cooled the concentrated phosphate buffer is added. The bovine albumin is gently dissolved in the prepared buffer solution. The pH is adjusted to 7.4 with 0.4 N NaOH. The solution is sterilized using Seitz filtration pads (Baird and Tatlock (London) Ltd) or membrane filters (Millipore (UK) Ltd) (pore size 0.22µm).

(*d*) L-*cystine* (BDH Chemicals Ltd). 0.8 g.

(*e*) *Inactivated rabbit serum*. 50 ml. Prepared as in 1(b) above. It has been found that adding inactivated rabbit serum enables maintenance of leptospira for longer periods.

(*f*) *Preparation*. 2800 ml of distilled water is added to a 5 litre flask and the following solutions added [numbers/letters in parentheses indicate solutions given in (a)–(e) above]: 160 ml of 25 × phosphate buffer stock solution (i); 200 ml of 20 × salts stock solution (ii); 4 ml copper sulphate stock solution (iii) and 40 ml of zinc sulphate stock solution (iv). Eighty millilitres of the ferrous sulphate solution (v) are added and the mixture shaken for 5 min. The medium will become hazy at this point, probably due to precipitation of insoluble phosphate. L-Cystine (d) (0.8 g) is added and the mixture shaken for a further 3 min. No attempt is made to completely dissolve the L-Cystine. The mixture is then filtered through a triple thickness of Whatman No. 1 filter paper (Scientific Supplies Co. Ltd.) The resulting solution should be clear.

The previously prepared Tween 80 solution (b) is added and the whole thoroughly mixed before adding a further 236 ml of distilled water. This base bulk solution is now autoclaved at 121°C for 15 min and allowed to cool.

When the base bulk solution has cooled, 1 litre of freshly prepared sterile bovine albumin solution (c) is added aseptically together with 80 ml of vitamin B_{12} working solution (vi,b), 0.4 ml of vitamin B_1 solution (vii) and 50 ml of sterile rabbit serum (e). The pH is checked and readjusted, if necessary, to pH 7.3–7.4. The complete medium is then sterilized by filtering through a 'sandwich' of membrane filters (pre-filter pad, 0.22, 0.45 and 1.2 µm pore-size membranes; Millipore Ltd) into either sterile glass tubes or bijoux bottles with plastic caps. The medium can also be dispensed into 100 ml amounts for use later.

The dispensed medium is then incubated at 37°C for 2 days and at 22°C for 3 days to ensure sterility.

Both media are very complicated to prepare, require considerable technical skill to manipulate and, in general, require subculture of the leptospira every 6–12 weeks.

C. Method

Isolates and stock cultures of strains for antigen preparation should be subcultured in either Korthof's medium or Ellinghausen and McCullough's medium with added 0.5% agar (semi-solid). Cultures should be incubated at 28°C until good growth is obtained. They should be examined by dark-field microscopy (× 40) for possible contamination and viability. Only those cultures demonstrating actively motile leptospira should be subcultured into further maintenance medium; those that are growing poorly should be transferred to liquid medium containing an increased concentration of rabbit serum (Turner, 1970) and reincubated at 28°C until profuse growth is seen. The whole subculturing procedure should be repeated every 12 weeks.

III. LIQUID NITROGEN

By far the easiest method of maintaining leptospira is storage in the vapour phase of liquid nitrogen. At the Leptospira Reference Unit (Hereford, UK), organisms suspended in Ellinghausen and McCullough's medium have been recovered from liquid nitrogen after storage for up to 2 years. However, recovery rates are low, usually about 10–20%, and the optimal conditions for long-term survival have yet to be established. Nevertheless, preserving leptospira in liquid nitrogen has been found easy to carry out and eliminates the need for continuous subculturing. With immediate subculturing into fresh medium following thawing, many of the primary functions of leptospira, such as pathogenicity and antigenic pattern, have been preserved.

A. Method

Leptospira are grown in the Ellinghausen and McCullough's medium described in Section II.B.2 until profuse growth is obtained. 0.5 ml of the culture is put into a 2 ml capacity plastic ampoule (Sterilin Instruments) and placed in the vapour phase of liquid nitrogen. When retrieval is required the ampoules are thawed quickly and the culture is immediately inoculated into Ellinghausen and McCullough's medium supplemented with an increased volume of inactivated rabbit serum (Section II.B.1.b) (20%). Subculture into fresh medium is usually required before full viability is re-established.

As mentioned previously, recovery rates using liquid nitrogen are low, usually about 10–20%.

IV. CONCLUSIONS

It is obvious that maintaining leptospira in a vital and active state is extremely difficult. Although some measure of success is possible using the methods described, there remains a need to establish alternative methods for maintaining these organisms. Future developments should re-examine freeze-drying, since this method of maintenance provides the greatest likelihood that leptospira keep their primary pathogenic and antigenic function intact.

BACTERIA

4.10. MAINTENANCE OF MICROORGANISMS BY SIMPLE METHODS

KHURSHEED AHMAD MALIK

DSM-Deutsche Sammlung von Mikroorganismen und
Zellkulturen GmbH, Braunschweig, FRG

I. Introduction 121
II. Methods 122
   A. Storage under liquid paraffin 122
   B. Storage in sterile distilled water 122
   C. Drying on silica gel 123
   D. Drying of cultures on glass or porcelain beads 125
   E. Liquid-drying of cultures using a simple apparatus .. 127
III. General notes 131
IV. Survival of preserved microorganisms 132

I. INTRODUCTION

The long-term preservation of cultures in small laboratories is often very difficult due to the absence of special equipment for deep-freezing or freeze-drying. The storage and preservation of cultures using simple methods is, however, possible in any laboratory (Antheunise, 1972; Lapage and Redway, 1974; Kirsop and Snell, 1984; Malik, 1985; Hill et al., 1990). In this chapter, the methods of storage under liquid paraffin, in distilled water and drying on glass or porcelain beads are described. Although the methods have proved to be safe, reliable and useful for the storage of a wide range of microorganisms, viability, shelf life and the stability of cultures is relatively poor when compared with preservation in liquid nitrogen and by liquid-drying (Malik, 1987c; Hill et al., 1990).

Cultures of several microorganisms that are sensitive to freezing or freeze-drying can successfully be preserved by liquid-drying, and a process is described here using simple apparatus that can easily be performed in most laboratories.

MAINTENANCE OF MICROORGANISMS 2nd Edn
ISBN 012–410351–0

II. METHODS

A. Storage under liquid paraffin

Living cultures can be maintained for longer periods of time by reducing the metabolic rate of the cells. This is achieved by using a medium with minimal nutrients, by creating a reduced gaseous environment or by storage at relatively low temperatures (4–10°C).

In this method living cultures are maintained in a reduced metabolic state by overlaying the cultures with sterile medicinal-grade liquid paraffin. The method has been successfully used for fungi, yeasts and some bacteria (Hartsell, 1956; Lapage and Redway, 1974). The average storage life for fungi is about 5–6 years and for bacteria about 2–4 years, depending on the strain.

1. Growth of micro-organisms

Cultures are grown on any suitable non-selective media (preferably in screw-cap tubes) until the late logarithmic phase of growth or until sporulation.

2. Procedure

Liquid paraffin (10 ml, medicinal grade) is poured into test tubes and is autoclaved. It is allowed to cool and is poured aseptically onto slants showing good growth. The paraffin layer should extend at least 2 cm above the upper end of the agar to prevent it drying out. The cotton plug of the culture tube is replaced with a sterile screw-cap and stored at 4–15°C.

3. Revival of cultures

To revive the cultures one loopful is taken out with an inoculation loop or needle and is touched aseptically onto sterile filter paper to drain off paraffin oil. The filter paper discs are sterilized by placing in glass Petri dishes and autoclaving. The oil-free inoculum on the loop is used to inoculate fresh agar medium (slant or plate). The remaining culture under the paraffin oil is stored for further use. The cultures are incubated at the appropriate growth temperature and are periodically examined for viability and purity.

B. Storage in sterile distilled water

This is a very simple and inexpensive method and is recommended for fungi, yeasts, plant pathogenic bacteria and actinomycetes. The average storage life is about 5 years (McGinnis *et al.*, 1974). Enterobacteriaceae show poor survival.

1. Growth of micro-organisms   Cultures may be grown on any suitable agar slants. In the case of sporulating fungi the cultures are incubated until sporulation is achieved. Other cultures are grown until the late logarithmic phase of growth.

2. Procedure   Distilled water (about 5 ml) is sterilized in screw-cap tubes. For use in the preservation of sporulating fungi, 0.1% Tween 80 is added to the water before autoclaving. Sterile distilled water (4–5 ml) is added to slants showing good growth and a thick cell suspension is prepared. In the case of fungi the spores are dislodged or the aerial growth is scraped off lightly using a sterile spatula or a strong inoculation needle.

  The suspension is transferred to the sterile screw cap tube, which is securely tightened to prevent water loss. The tubes are stored at 5–15°C.

3. Revival of cultures   For revival, the tube is shaken and 0.2 ml of the suspension is transferred onto fresh medium (slant or a plate). The remaining suspension can be stored for further use. The cultures are incubated at the appropriate growth temperature and are periodically examined for viability and purity.

C. Drying on silica gel

  This is a very economical method that is valuable for laboratories with limited resources. The cultures are dried on crystals of silica gel (Fig. 1). For recovery, only a few crystals bearing the dried cultures are removed, the bulk of the stock culture remaining stored for future use. Although many microorganisms (filamentous fungi, yeasts, bacteria, mycoplasmas) can be successfully preserved, a number of species may not survive (Perkins, 1962), and cultures should not be preserved solely by this method without previous experimentation to establish likely survival levels. See Chapters 5 and 6, this volume, for filamentous fungi and yeasts, respectively.

  Care should be taken that the heat liberated during rehydration does not damage the cells. Anhydrous silica gel and all media and glassware must be thoroughly chilled before use. Due to toxicity, blue silica gel containing an indicator ($CoCl_2$) should be avoided.

1. Growth of micro-organisms   Cultures are grown on appropriate growth media until the late logarithmic phase of growth is reached or until sporulation occurs.

2. Preparation of suspending medium   Protective agents are used to suspend cells to be dried in order to protect against drying injury. A suitable suspending medium is 10% (w/v) solution of skimmed milk (Bacto-Skim milk, Difco,

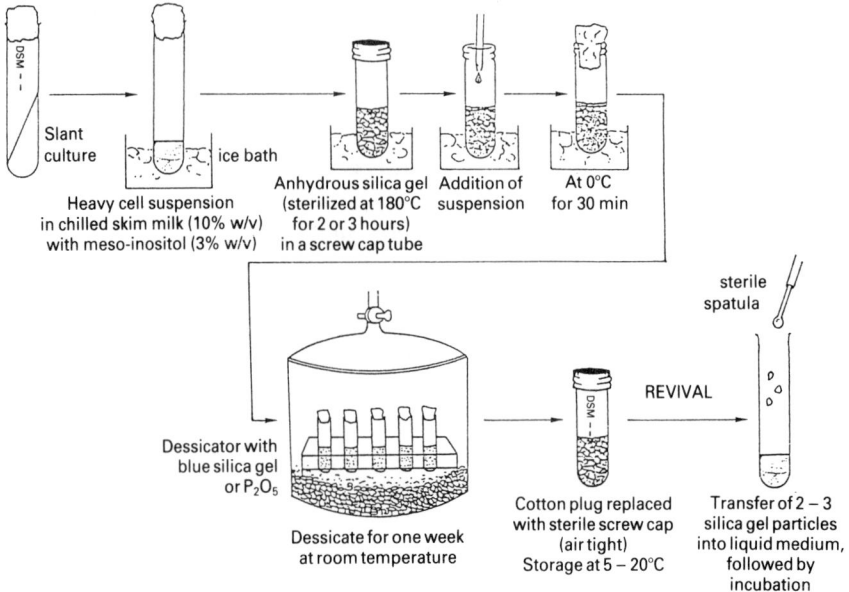

Fig. 1   Drying cultures on silica gel.

0032) containing 3% (w/v) *meso*-inositol or 5% (w/v) honey. It is sterilized at 115°C for 13 min.

3. Procedure   Screw-cap tubes containing 5 g of anhydrous silica gel (6–22 mesh, without indicator) are prepared. These are sterilized (dry heat) at about 180°C for 1–2 h and are closed tightly when hot in order to prevent entry of moist air. The rubber lined caps of the tubes are autoclaved separately then dried in an oven at 70–80°C. Before use, the tubes containing dry silica gel are chilled in an ice bath.

A thick cell suspension (containing at least 10^8 cells ml^{-1}) of the culture is prepared in the suspending medium. The suspension is cooled in an ice bath and about 0.5 ml is added drop by drop to the pre-chilled silica gel. The tubes are shaken to distribute the cells over the crystals of silica gel. The screw cap is replaced with a sterile cotton plug. The tubes are placed in a desiccator containing blue indicator silica gel or P_2O_5 for about 1 week at room temperature. After this time, the cotton plug is replaced again with a sterile screw cap and it is tightened. The tubes are stored preferably at 5–10°C.

4. Revival of cultures   For reactivation, a few crystals of silica gel are transferred with a sterile spatula or a loop into a suitable liquid growth medium. The remaining crystals are stored for further use. It is important to transfer the tubes first to room temperature before opening to prevent condensation of water in the tubes. Cultures are first incubated at a temperature lower than the optimal growth temperature. A week or more may be required for good growth. As with other preservation methods the culture is examined at intervals for viability and purity after reactivation.

D. Drying of cultures on glass or porcelain beads

Cultures of several microorganisms that are sensitive to freezing or freeze-drying, can successfully be preserved by drying under more gentle conditions. In simple drying procedures a small volume of cell suspension is spread over a large surface area. In this method (Fig. 2) cells are spread over the surface of glass or

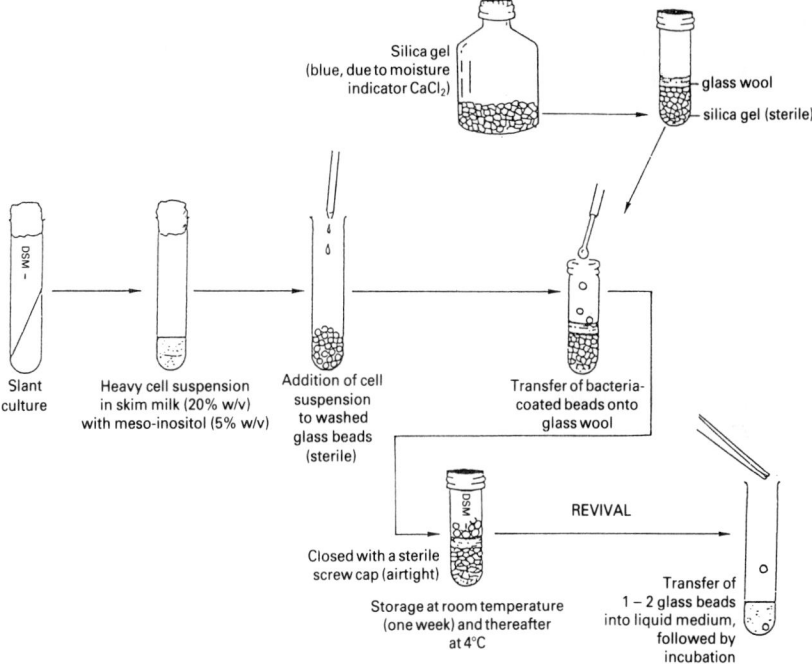

Fig. 2   Drying cultures on porcelain beads.

porcelain beads and gentle drying is achieved over a desiccant without the application of vacuum. It is a very useful method for spore forming fungi and bacteria (Hunt *et al.*, 1958). The average storage period for spore forming microorganisms is more than 10 years. With the use of a protective agent, such as milk, honey, sugars or serum, even non-spore forming bacteria can be successfully maintained for many years.

1. Preparation of beads   Glass or porcelain beads of 2–5 mm diameter are washed in tap water with a mild detergent. The beads are rinsed with dilute

HCl to neutralize alkalinity. They are then washed several times in tap water until the pH of the wash water is the same as that of tap water. The beads are finally washed in distilled water, then dried in an oven. These are placed in screw-cap tubes and are sterilized by dry heat at about 180°C for 1–2 h. The screw-caps are tightened when hot in order to prevent entry of moist air. The rubber-lined caps for the tubes are autoclaved separately, then dried in an oven at 70–80°C. The glass beads can be obtained from laboratory glassware suppliers or embroidery beads, which are commercially available, can be used. Different coloured beads can be used to differentiate various groups of bacteria during storage.

2. Preparation of storage tubes

A screw-cap tube is half filled with blue indicator silica gel and covered with a 1 cm layer of glass wool (Fig. 2). It is dry-heat sterilized at 180°C for 2 h. During dry-heat sterilization care should be taken that no rubber septa or gaskets are present in the metallic screw caps. The tubes are closed tight when hot, in order to prevent entry of moist air. They are labelled for each organism.

3. Preparation of suspending medium

Skimmed milk (20% w/v in H_2O) containing 5% *meso*-inositol (w/v) is sterilized in Universal bottles at 115°C for 13 min.

4. Growth of micro-organisms

Cultures are grown on appropriate growth media until the late logarithmic phase of growth or until sporulation occurs.

5. Procedure

About 1 ml of sterile suspending medium is pipetted into a glass tube.
   A thick cell suspension (at least 10^8 cells ml^{-1}) is prepared by transferring cell material from the growth slant into the suspending medium. Using a wire loop, the growth is emulsified and well mixed to make a homogeneous suspension. The beads (about 20–50, depending upon their size) are transferred aseptically to the bacterial suspension. The suspension is aspirated several times to ensure that the air bubbles around and (in the case of perforated beads) inside the beads are displaced by the bacterial suspension. After the beads are thoroughly wetted, the excess suspension is drained off from the bottom of the vial using a sterile Pasteur pipette or a syringe. The beads, coated or impregnated with the cells, are transferred aseptically into the storage tubes. The screw cap is replaced and tightened securely and kept at room temperature for about 1 week. Within a few days the cultures on the beads will dry out and the upper layer of the silica gel will turn pink through hydration. The remaining blue silica gel will ensure desiccation during storage at 4–10°C. If too much moisture is transferred with the beads or if the screw cap tube is

not completely airtight, the blue silica gel becomes pink or colourless within a day or so. The perforated embroidery beads take a longer time for drying.

6. Revival of cultures   For reactivation one or two beads are transferred aseptically into liquid culture medium. Occasionally the beads stick together but can easily be separated by a sterile spatula, forceps or tweezers. The remaining beads are stored for further use. The culture is examined for viability, purity and retention of particular characteristics after reactivation.

E. Liquid-drying of cultures using a simple apparatus

Liquid-drying (L-drying) involves vacuum drying of samples from the liquid state without freezing, as described for industrial and marine bacteria in chapter 4.4. Several reports describe the successful application of this method (Annear, 1956, 1958; Lapage and Redway, 1974; Banno and Sakane, 1979; Hieda, 1981). Specialized equipment is generally required to create the conditions conducive to the liquid-drying process, but a procedure is described here using a simple apparatus (Fig. 3), and based on a method which has recently been published for the successful preservation of sensitive microorganisms (Malik, 1990b).

Protective media such as skimmed milk and *meso*-inositol, or honey, glutamate or raffinose (Malik, 1976, 1988a), may be used as a suspending medium in order to protect the cells against drying injury. Several species of anaerobic bacteria, which are sensitive to aerobic L-drying, can successfully be preserved using activated charcoal (5% w/v) in the suspending medium, together with any of the above protective agents (Malik, 1990a).

The method described below has been successfully applied to the preservation of a number of sensitive microorganisms which otherwise are damaged by freezing or freeze-drying and fail to survive. It uses a double vial method in which the ampoules containing the liquid-dried cell material are sealed under vacuum in a soft glass tube. Throughout the procedure the temperature of the cell material is maintained at 20°C.

1. Preparation of carrier material   Ampoules of neutral glass (45 × 10 mm) are filled with 0.1 ml of 20% (w/v) skimmed milk (Bacto, Difco 0032) containing 10% (w/v) neutral activated charcoal and one of the effective protective agents such as 5% (w/v) *meso*-inositol, 5% (w/v) glutamate, 5% (w/v) raffinose or 10% (w/v) honey. The activated charcoal is of medicinal grade. It is available from Caelo, 4010 Hilden, FRG, but any bacteriological activated charcoal of comparable quality can also be used. The ampoules are loosely plugged with

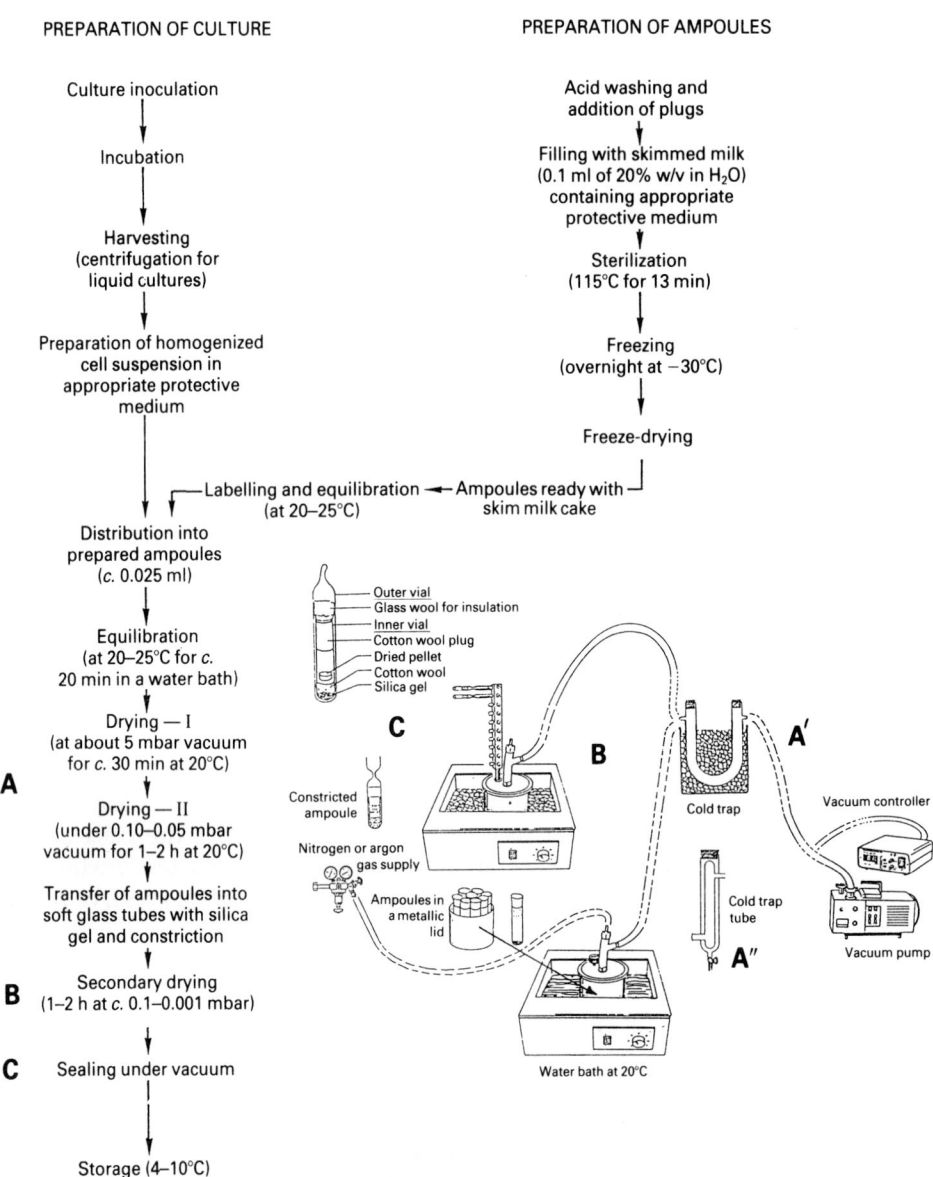

PREPARATION OF CULTURE

Culture inoculation

Incubation

Harvesting
(centrifugation for
liquid cultures)

Preparation of homogenized
cell suspension in
appropriate protective
medium

Distribution into
prepared ampoules
(c. 0.025 ml)

Equilibration
(at 20–25°C for c.
20 min in a water bath)

Drying — I
(at about 5 mbar vacuum
for c. 30 min at 20°C)

A

Drying — II
(under 0.10–0.05 mbar
vacuum for 1–2 h at 20°C)

Transfer of ampoules into
soft glass tubes with silica
gel and constriction

B   Secondary drying
(1–2 h at c. 0.1–0.001 mbar)

C   Sealing under vacuum

Storage (4–10°C)

PREPARATION OF AMPOULES

Acid washing and
addition of plugs

Filling with skimmed milk
(0.1 ml of 20% w/v in H$_2$O)
containing appropriate
protective medium

Sterilization
(115°C for 13 min)

Freezing
(overnight at −30°C)

Freeze-drying

Labelling and equilibration ← Ampoules ready with
(at 20–25°C)   skim milk cake

Outer vial
Glass wool for insulation
Inner vial
Cotton wool plug
Dried pellet
Cotton wool
Silica gel

C

Constricted
ampoule

Nitrogen or argon
gas supply

Ampoules in
a metallic
lid

B

Cold trap

Cold trap
tube

Water bath at 20°C

A′

Vacuum controller

A″

Vacuum pump

Fig. 3   L-drying process using simple apparatus.

non-absorbent cotton wool and sterilized at 115°C for 13 min. These are frozen at about −30°C for a few hours and then freeze-dried in bulk for about 6 h using a standard freeze-drying technique. The ampoules can also be liquid-dried using a simple apparatus as described under the liquid-drying procedure. Large batches of ampoules should be avoided if the capacity of the system (vacuum pump, condensing temperature of the cold trap, etc.) is inadequate to remove the evaporated liquid.

2. Preparation of protective agents

Solutions of protective agents like *meso*-inositol (5% w/v), honey (10% w/v), sodium glutamate (5% w/v), raffinose (5% w/v) are prepared in distilled water, filter-sterilized and stored at 4°C. For oxygen-sensitive microorganisms, 10% (w/v) neutral activated charcoal with 10% (w/v) *meso*-inositol is prepared in a screw-cap bottle (Malik, 1990a,b). The solution is boiled, bubbled with N_2 gas and the bottle is closed tightly and autoclaved at 115°C for 13 min.

3. Preparation of cell suspension

The cultures are grown on appropriate media. A thick cell suspension (at least 10^8 cells ml^{-1}) is prepared in a suitable protective medium. In the case of liquid cultures, the cells are harvested by centrifugation for 30 min at 4000 *g* in screw-cap bottles and the pellet is suspended in a protective medium to yield a heavy cell suspension.

4. Filling ampoules with cell suspension

The prepared ampoules, containing a thin disc or plug of carrier material, are held at 20°C to equilibrate for a few minutes (see Malik, 1990b). To each ampoule about 0.025 ml (one drop from a Pasteur pipette) of cell suspension is carefully delivered onto the carrier material so as not to touch the sides of the ampoules. The ampoules are quickly placed in small metal containers or lids (aluminium caps often used in laboratories are also suitable) and transferred into a metallic jar maintained at 20°C in a water bath (see Fig. 3A). After allowing to stand for a few minutes to allow adsorption of the cells onto the carrier material and temperature equilibration, the material is dried under vacuum.

For strains that are very sensitive to oxygen, the prepared ampoules are first placed for a few hours in Bio-bags (Type A, Marion Scientific Corporation) or Anaerocult P bags (E Merck) to ensure complete anaerobiosis. Ampoules are removed one at a time, inoculated with the reduced cell suspension (see details in ch. 4.7, page 91) and transferred immediately to fresh anaerobic bags. The bags are held for about 1 h at room temperature for temperature equilibration and removal of oxygen. The ampoules are then subjected to drying under vacuum.

5. Preparation of coolant and the L-drying procedure

The outline of the L-drying procedure and the major steps involved are shown in Figure 3. In the double vial system, drying is carried out in two stages involving primary and secondary drying. The primary drying is also achieved in two stages.

The cold trap is chilled to about −35°C and is connected to the vacuum pump and the metallic evacuation jar (maintained at 20°C in a water bath). A vacuum controller is attached between the cold trap and the vacuum pump. The cold trap may be made either from U-shaped glass tubing, or preferably metal tubing of about 2–3 cm diameter and 30 cm in length. It is filled with blue indicator silica gel and is placed in a metallic beaker as shown in Fig. 3A' and A". It is chilled to −35°C.

The cooling mixture for the trap is prepared by placing a mixture of ethylene glycol:water (1:1) in a deep freezer and cooling to below −30C, but preferably to −40C. The cooling mixture is poured into the cold trap, filling it to the maximum level. In addition, some of the cooling mixture is sealed in deep freeze plastic bags and cooled in a deep freezer or over liquid nitrogen. This super cooled coolant is added to the cold trap throughout the drying procedure to keep the temperature as low as possible, the bags being changed every 20–25 min.

Commercially available anti-freeze liquids used in car radiators are also satisfactory coolants. Alternatively, a double jacketed straight tube (exterior 6 × 30 cm, interior 3 × 25 cm, with outlet and inlet tubes of 2 × 5 cm, as shown in Fig. 3A") can be used as a cold trap. Another alternative is to use dry ice for freezing the material in the ampoules and for adding to the cold trap throughout the procedure.

First stage drying (Drying I in Fig. 3A) is carried out for 30 min at 5–10 mbar. After this time the vacuum controller is adjusted to between 1.0 to 0.1 mbar vacuum and the second stage of drying (Drying II in Fig. 3A) is carried out for about 1 h. Upon completion, the vacuum is replaced by nitrogen gas. This is important for strict anaerobes and for ampoules which are to be stored without sealing under vacuum (see General Notes).

For secondary drying and sealing, the ampoules are transferred to soft glass tubes (130 × 12 mm) containing silica gel and cotton plugs. The outer tubes are then constricted by hand or using an Edwards' Ampoule Constrictor. They are attached to a manifold and mounted on an evacuation jar (Figure 3B). Secondary drying is carried out for 1–2 h at 0.1–0.001 mbar. The pink silica gel will return to a blue colour when the process is complete. The constricted outer tubes are carefully sealed by hand or using a Flaminaire blow torch (Fig. 3C).

6. Revival of cultures

The contents of the liquid-dried ampoule are resuspended in sterile liquid growth medium (pre-reduced in the case of anaerobes) and transferred (or injected in the case of anaerobes) into 15–20 ml of growth medium. Freshly inoculated phototrophic cultures are placed for a few hours in the dark at appropriate incubation temperatures and later transferred to normal growth conditions in the light. The dried cultures are incubated at a slightly lower temperature than the optimum growth temperature, and for longer periods.

7. Estimation of viability

Survival levels are checked before L-drying, immediately after L-drying and after storage. For the estimation of viability, serial dilutions are prepared in appropriate liquid media. From each serial dilution 0.1 ml volumes are plated on agar plates. The number of colonies are counted from the plates and average colony forming units per sample are calculated. The revived cultures are also observed for mutation, change in colony morphology or other characters. For cultures that are difficult to grow in or on agar, only liquid dilution series are made. In such cases, the number of cells is determined using the most probable number method (MPN).

8. Long-term storage   As storage at low temperatures will extend the shelf life of cultures, liquid-dried cultures are stored at about 5°C. If L-dried cultures are stored unsealed, they may be safely stored at −30°C for several years. The author has found that survival of such cultures compares well with the survival of freeze-dried cultures sealed under vacuum and stored at 9°C (Malik, 1976).

III. GENERAL NOTES

Sterilization of materials and media is carried out at 121°C for 15 min unless indicated otherwise.

Cultures to be preserved are best grown on suitable non-selective media until the late logarithmic stage of growth; however, cultures that have begun to lyse should not be used. Growth in rich media should not be used as in some cases it may result in the accumulation of toxic end-products which cause stationary phase cells to become non-viable. The use of a solid non-selective medium aids the detection of contaminants.

For halophiles, the protective agent should be prepared in appropriate growth medium instead of water, or the suspending medium should be supplemented with an appropriate amount of salt.

A high level of residual moisture or exposure to oxygen may have detrimental effects on the dried product. The dried cell material is hygroscopic and its exposure to moisture during storage can destabilize the product. The higher the storage temperature, the faster a product will degrade. Thus, the storage of dried cultures at lower temperatures will extend the shelf life. Similarly, sealing the dried product under a vacuum prolongs the shelf life.

Improvement in viability can be achieved through the use of different reactivation procedures. In the case of some sensitive microorganisms, cultures show lower survival levels on agar medium than in liquid medium. Again, lower survival levels are obtained on agar medium of higher surface tension (such as nutrient agar) than on mineral medium of somewhat lower surface tension. The use of activated charcoal in the rehydration medium for reactivation has proved useful (Malik, 1990a,b). The author has found that the presence of activated charcoal (0.01 to 0.1%) in medium improves the stability and viability of sensitive microorganisms during long-term storage as living cultures.

Higher viability counts have been obtained following incubation at lower temperatures than the normal optimum growth temperatures. A few species may exhibit a prolonged lag period and must be incubated for longer periods. Normal growth rates usually are restored after a second transfer into fresh medium.

It is important to obtain the highest recovery levels possible from the preserved material as this minimizes the possibility of the selection of cells resistant to the technical manipulations employed during preservation.

IV. SURVIVAL OF PRESERVED MICROORGANISMS

Although successful preservation of a wide range of microorganisms is possible using the simple methods described in this chapter, lower viability and stability levels have been observed in some strains compared with those obtained using more sophisticated methods. The methods should not be used for some sensitive strains (Table I) nor for genetically engineered microorganisms or plasmid-bearing strains.

Experience has shown that large-celled Gram-negative bacteria, bulky poly-β-hydroxy butyric acid-filled cells, large *Spirillum*-like cells and temporarily enlarged cells are sensitive to preservation stress and have poor survival during long-term maintenance (Malik, 1988a). Again, isolates from surface water are less resistant to preservation than soil isolates. By contrast, cells with spores or cysts, smaller cells or those harvested from the late logarithmic growth phase survive better with simple drying methods.

Table I: Examples of microorganisms that do not survive well following preservation by simple methods. It should be noted that many may be preserved successfully by liquid-drying

Aquaspirillum
Borrelia
Campylobacter
Hydrogenophaga
Neisseria
Spirillaceae
Vibrios
Zymophilus spp
Many plasmid-bearing *Pseudomonas* spp. and *Alcaligenes* spp.
Anaerobic and methanogenic bacteria
Many nitrogen-fixing bacteria such as Azotobacteriaceae and *Azospirillum*
Anoxygenic phototrophic bacteria such as Rhodospirillaceae (brown), Chlorobiaceae, Chromatiaceae, *Chloroflexus* and *Heliobacterium*
Kluyveromyces spp., *Saccharomyces cerevisiae*, *Leucosporidium* spp., *Sporobolomyces* spp. and some high lipid-producing yeasts
Filamentous fungi such as *Phytophthora* spp. and *Pythium* spp.
Unicellular algae such as *Dunaliella* and *Chlamydomonas*

FILAMENTOUS FUNGI

5. MAINTENANCE OF FILAMENTOUS FUNGI

D. SMITH

International Mycological Institute
Kew, London, UK

I.	Introduction	134
II.	Maintenance by subculturing	135
	A. On agar	135
	B. On agar in the refrigerator	136
	C. On agar under oil	136
	D. On agar blocks in water	136
III.	Maintenance by drying	137
	A. On anhydrous silica gel	137
	B. In soil/sand	138
IV.	Maintenance by freeze-drying	138
V.	Maintenance by freezing	139
	A. On agar slopes at −20°C	139
	B. In liquid nitrogen	140
VI.	Storage and survival by subculture	141
	A. On agar at growth temperatures	141
	B. On agar in the refrigerator	142
	C. On agar under oil	143
	D. On agar blocks in water	146
VII.	Storage and survival of dried cultures	146
	A. On anhydrous silica gel	146
	B. In soil/sand	150
VIII.	Storage and survival of freeze-dried cultures	151
IX.	Storage and survival of frozen cultures	152
	A. On agar slopes at −20°C	152
	B. In liquid nitrogen	155
X.	Black light: its use to induce sporulation in fungal cultures	157
XI.	Mite prevention	158
	A. Hygiene	158
	B. Fumigation	159
	C. Mechanical and chemical barriers	159
	D. Cold storage	159

MAINTENANCE OF MICROORGANISMS 2nd Edn
ISBN 012–410351–0

I. INTRODUCTION

In the past, collections of fungi were kept by serial transfers from depleted to fresh media. This method is still used for working collections holding small numbers of cultures that are in constant use. However, this is a very time-consuming and labour-intensive exercise, especially when a large number of organisms is kept. Other problems associated with this method are that the cultures are often short-lived and are liable to contamination both by other microorganisms and by mites. Moreover, their morphology or physiology may change due to adaptation to growth on artificial media; for example, they may lose their ability to sporulate or reproduce sexually, or lose certain physiological properties. Again, plant pathogenic fungi grown on media for several generations often lose their pathogenicity and must be transferred to their host to check whether this has occurred.

Over the years, several methods of preservation have been developed to eliminate problems encountered with serial transfers. One of the aims has been to extend the storage period between subculturing by allowing growth to continue at a reduced rate. This has been achieved by growing the fungus on a weak medium, by lowering the incubation temperature, or by growing under conditions that prevent dehydration and provide reduced oxygen tension. The latter can be accomplished by the addition of liquid paraffin to a culture growing on standard media at normal growth temperatures. Most of these techniques are still open to many of the problems encountered during normal serial transfers.

The methods that have proved most successful have been those that reduce metabolism to an extent that induces artificial dormancy. This is usually achieved by dehydration or freezing. Freeze-drying (lyophilization) allows the dehydration of fungi to a level that halts metabolism. Dehydration is achieved by the sublimation of ice and is continued until a residual moisture content of between 1 and 2% is achieved. To halt metabolism by freezing, the organism must be subjected to temperatures below $-130°C$ (Mazur, 1966). Liquid nitrogen at a temperature of $-196°C$ has proved very useful as a coolant and storage medium for organisms. Liquid air has also been used (Joshi et al., 1974), though the risks involved are much greater because of potential explosion.

The choice of preservation method depends upon many factors. For example, the preservation requirements of a small teaching collection differ from those of a national depository. Again, taxonomic collections require methods that stabilize the morphology and physiology of the organism. Similarly, industrial collections place much emphasis on techniques that maintain genetic stability, especially for strains that have special features required for industrial processes, such as the production of antibiotics and enzymes. Above all, the availability of facilities and funds can be a key factor in the choice of a preservation technique.

Many reviews cover the preservation techniques that are available for microorganisms (Fennell, 1960; Nei, 1964; Onions, 1971; Jong, 1978; Heckly, 1978; Smith, 1988), particularly the techniques of freeze-drying (Raper and Alexander, 1945; Last et al., 1969; Rowe, 1971; Smith, 1983b) and liquid nitrogen storage (Hwang, 1960; Butterfield et al., 1974). Genetic stability can be achieved by freeze-drying (Onions, 1971; Heckly, 1978) and liquid nitrogen storage (Prescott and Kernkamp, 1971), both of which prove to be excellent long-term storage techniques. However, for the successful use of any of the methods it is essential that the culture is in good condition. Optimum growth and sporulation will lead to better survivals after preservation; poor cultures are rarely improved by preservation.

II.   MAINTENANCE BY SUBCULTURING

A. On agar

Fungi can be grown on a wide range of agar media (Booth, 1971a). It is common practice to attempt to simulate the natural conditions of an isolate in order to obtain its fullest development. In many cases the relationship between the organism and others found in its natural environment can not be simulated in pure culture. However, a natural substrate can be provided and the use of sterile natural media or vegetable extracts may be helpful (Dade, 1960). Some fungi have special requirements for successful storage and, wherever possible, these should be provided. Dermatophytes survive best on hair (Doory, 1968), some water moulds are best storë in water, often with the addition of plant material (Goldie-Smith, 1956), and other more sensitive water moulds may require aeration (Clark and Dick, 1974; Webster and Davey, 1976). It may be necessary to grow fungi in mixed culture; *Dictyostelium*, for example, may require bacteria, and the mycoparasite *Piptocephalis* can be provided with its microfungus host, for example *Cokeromyces* or *Penicillium*.

The culture vessel may be a Petri dish, a McCartney bottle, a plastic or glass universal bottle, or a test tube. In selecting a container, various aspects of storage must be taken into consideration and storage space is perhaps the first selection criterion.

The material from which the vessel is made is important if the isolate requires 'black light' treatment (see Section X), in which case plastic should be selected in preference to glass.

The type of inoculum can affect the quality of the fungal collection. It is thought that a mass spore transfer is by far the best technique and that mycelium transfer should only be used where no spores are available (Onions, 1971). There are occasions when this is not the case. Continual transfer of the spores of ageing Mucorales can lead to the deterioration of the culture and the transfer of spores together with mycelium and substrate will avoid this problem. When mycelium transfer must be made it should be done from the growing edge of the colony. Pythiaceae are best subcultured by removal of the basal felt (Onions, 1971).

Single-spore cultures (Gordon, 1952) may be used with some unstable strains where, with experience, the wild-type spore can be selected and transferred to fresh medium. This technique is also useful when separating an isolate from a more rapidly growing contaminant.

B. On agar in the refrigerator

The fungi are subcultured (see Section II.A) and stored in a refrigerator at a temperature between 4 and 7°C.

C. On agar under oil

Healthy, mature cultures grown on slants of suitable agar in 30 ml universal bottles without cap liners are covered with twice autoclaved (121°C for 15 min) liquid paraffin (medicinal paraffin, specific gravity 0.830–0.890) to a depth of 1 cm above the highest point of the agar slope or fungal growth. The bottles, with their caps loose, can be stored at room temperature or in an air-conditioned room maintained at 15–20°C.

Retrieval of a culture maintained in mineral oil is achieved by removing a small amount of the fungal colony on a mounted needle and draining away as much oil as possible on the neck of the culture bottle. The fungus is then streaked onto a suitable agar medium. More than one subculture may be necessary after retrieval as the growth rate can very often remain slow because of adhering oil. Better results may be obtained by inoculating mid way down an agar slope and allowing oil to drain and the fungus to grow up the slope away from the oil.

D. On agar blocks in water

The fungi to be stored are grown on suitable agar media in Petri dishes. Agar blocks, 6 mm^3, are cut from the growing edge of the fungal colony and transferred to sterile distilled water in McCartney bottles. The lids are tightly screwed down and the bottles stored at room temperature. Other methods only require harvested spores to be suspended in water (Castellani, 1939, 1967).

III. MAINTENANCE BY DRYING

A. On anhydrous silica gel

1. Preparation of silica gel

Glass universal bottles are one-quarter filled with purified non-indicator silica gel, 6–22 mesh, and sterilized in an oven at 180°C for 2–3 h. The sterile gel is stored at 37°C, or in a desiccator, to avoid absorption of moisture; if this occurs the gel must be re-sterilized before use. The bottles are placed in trays of water to a depth above the level of the silica gel. The water is frozen by placing the trays in a deep-freeze (-17 to -24°C).

2. Suspending medium

A 5% (w/v) solution of non-fat skimmed milk is prepared in distilled water and dispensed into universal containers in 5 ml amounts and autoclaved immediately at 115°C for 10 min. The skimmed milk solution is cooled to 4°C and stored at this temperature prior to use.

3. Spore suspension

Sporulating cultures of the fungi are placed in the refrigerator and cooled. The cooled skimmed milk is added to the cooled culture and a spore suspension is prepared by gently scraping the fungal colony with a sterile wire or glass rod.

4. Dispensing and drying the fungal suspension

The spore suspension is added to the cooled silica gel crystals in the trays of frozen water using a Pasteur pipette. Only three-quarters of the gel should be wetted to avoid over-saturation. The gel bottles are left in the ice bath for about 20 min until the ice around them has melted a little. The crystals are agitated to ensure thorough dispersion of the suspension. The bottles are kept at 25°C until the crystals readily separate when shaken, which is usually between one and two weeks.

5. Viability

The viability is checked by sprinkling a few crystals onto a suitable medium and assessing the amount of growth from each crystal. If growth is satisfactory, the caps of the universal bottles are screwed down tightly and stored over indicator silica gel in an airtight container, such as plastic freezer boxes with good seals, at 4°C. The indicator gel will require replacing once or twice a year.

The 'spent' gel can be dried out by heating at 180°C for 2 h or until the colour is fully restored.

B. In soil/sand

1. The soil

Garden loam, having a water content of approximately 20%, is put into glass universal bottles filling to between half and two-thirds capacity. This is sterilized by autoclaving at least twice at 121°C for 20 min, allowing the soil to cool between each autoclaving. Sand can be similarly sterilized and used instead of soil (Mehtra *et al.*, 1977).

2. The inoculum

A spore suspension is prepared by adding 5 ml of sterile distilled water to a culture and gently scraping the colony to release the spores. If the isolate does not sporulate in culture, a suspension of mycelium can be used.

3. Procedure

The suspension is dispersed in 1 ml amounts into the sterile soil or sand and a growth or drying-out period is allowed at room temperature. Two to three days is usually long enough for fast-growing isolates, but for most other isolates a period of 10–14 days allows the fungi to utilize the available moisture and allows growth to slow down sufficiently for storage.

4. Storage and recovery

The soil/sand culture bottles are stored with loose caps in the refrigerator (4–7°C). The fungi can be revived by sprinkling a few grains of soil onto a suitable medium.

IV.   MAINTENANCE BY FREEZE-DRYING

1. Preparation of ampoules

The isolate number and date of processing are printed on the outside of (0.5 ml) glass ampoules (Anchor Glass Co. Ltd) using a reverse type printer (Rejafix Ltd). The ampoules are covered with lint caps and heat-sterilized at 180°C for 3 h. The sterilization will also dry and fix the ink.

2. Inoculum

A spore suspension of the isolate is prepared by adding 10 ml of a 10% (w/v) skimmed milk and 5% (w/v) inositol mixture (autoclaved at 115°C for 10 min) to a mature sporulating culture grown on a suitable growth medium. The fungal colony is then scraped to release the spores.

3. Inoculation of ampoules

Approximately 0.2 ml of the spore suspension is added aseptically to each sterile ampoule using a Pasteur pipette. The filled ampoules are loaded into the centrifuge racks of a Modulyo or Super Modulyo spin freeze-dryer (Edwards High Vacuum Ltd).

4. Freeze-drying procedure

(1) The loaded centrifuge racks are placed in the spin-freeze accessory of the Modulyo or Super Modulyo freeze-dryer and the chamber secured. The centrifuge is switched on and the ampoules spun. The chamber is evacuated and when a pressure lower than 0.5 Torr is achieved the product will be frozen and the centrifuge is switched off. Drying is carried out for a further 3 h.

(2) The chamber is brought to atmospheric pressure. The ampoules are removed and plugged with sterile cotton wool which is compressed to a depth of 1 cm and pushed down inside the ampoule until it is just above the slope of the dried suspension. The ampoules are then constricted using a Flame Master air–gas torch (Buck & Hickman) about 1 cm above the top of the cotton-wool plug.

(3) The constricted ampoules are attached to the manifold of the secondary drying accessory of the freeze-drying unit and evacuated over phosphorus pentoxide desiccant. The interval during which the freeze-dried material is exposed to oxygen and water vapour in the atmosphere must be kept to a minimum as over exposure can cause deterioration (Rey, 1977). The drying is usually carried out overnight for 17 h leaving a residual moisture content of between 1 and 2%.

(4) The ampoules are sealed under vacuum at the point of constriction using a crossfire burner [Wesley Coe (Wingent) Ltd; Adelphi Manufacturing]. The ampoules are stored in an air-conditioned room maintained at between 15 and 20°C. After 3–4 days storage an ampoule is opened and the fungus reconstituted by the addition of 3–4 drops of sterile distilled water from a Pasteur pipette and allowing at least 15–20 min for absorption of moisture by the spores. The contents of the ampoule are streaked onto a suitable agar medium and successful retrieval is determined by resumption of normal growth and sporulation.

V. MAINTENANCE BY FREEZING

A. On agar slopes at −20°C

The cultures are prepared as for normal transfer (see Section II.A) and when mature they are placed in a deep-freezer at −20°C. The culture is recovered by removing a small portion of the frozen colony, placing it on a suitable agar medium and allowing it to thaw at room temperature. The frozen stock culture is replaced in the freezer as quickly as possible to avoid thawing.

B. In liquid nitrogen

1. Preparation of ampoules

The strain number is written, using a permanent felt tip pen (Scotts Office Equipment Ltd), or printed, using a reverse type printer (Rejafix Ltd), on to 2.0 ml borosilicate glass ampoules (Anchor Glass Co. Ltd). The ampoules are covered with individual tin-foil caps and sterilized in an oven at 180°C for 3 h. Screw-capped 2 ml polypropylene cryotubes (Lab M) can also be used for storage in the vapour phase.

2. Culture preparation

The fungi are grown on slopes of suitable agar in 30 ml universal containers (Sterilin Instruments). After growth is well established the cultures are transferred to a refrigerator at 4–7°C to allow continuing growth and a degree of cold hardening. Not all fungi will grow at these temperatures, and for such isolates the cold-hardening stage may be entirely omitted or may be reduced to a short storage period prior to freezing. This stage is isolate-dependent and can be determined by pre-growth tests before storage in liquid nitrogen.

3. Preparation of inoculum

Ten millilitres of 10% (v/v) glycerol are dispensed into glass universal bottles and autoclaved at 121°C for 15 min, cooled and stored at room temperature prior to use. The glycerol is added to the culture and the spores suspended by gentle agitation and scraping of the fungal colony. Where only mycelium is present, agar plugs can be taken using a cork borer or agar blocks cut with adhering colony; these may be added to 0.5 ml of the glycerol in the cryotubes.

4. Inoculation, sealing, and testing of ampoules

The fungal inoculum is added in 0.5 ml amounts to the sterile borosilicate glass ampoules which are heat-sealed using a Flame-Master air–gas torch. The seals are tested by immersing the ampoules in an erythrosin B dye bath at 4–7°C, which also pre-cools the fungal suspensions.

5. Freezing protocol

The ampoules containing the fungal suspension are frozen at a suitable cooling rate determined prior to storage. In many cases 1°C per minute will suffice; this can be achieved by suspension of the ampoules in the vapour phase of a liquid nitrogen refrigerator at −35°C for 40–45 min. This is followed by plunging the ampoules into liquid nitrogen and cooling them to −196°C. Alternatively, the cryotubes or glass ampoules can be placed in the chamber of a controlled freezer (Planer Products Ltd) and cooled at the most suitable rate. Optimum cooling rates for the strains to be preserved should be pre-determined to get the best results.

6. *Storage and viability*

The ampoules are stored either clipped to aluminium canes in boxes in a 250 litre refrigerator (Union Carbide) or in the drawer rack system of a wide neck 320 litre (Union Carbide) refrigerator in contact with the liquid nitrogen to retain low temperatures. To check the viability of cultures the ampoules are removed from the liquid nitrogen refrigerator and thawed by agitation in a water bath at 37°C. The temperature of the suspension must not reach 37°C, as this could be detrimental. When the suspension has thawed, it is streaked onto suitable agar medium, incubated and examined for growth.

7. *Special treatments*

Some fungi are damaged by excessive manipulation. To avoid this, they are placed in the ampoules on small slivers of agar. Alternatively, they may be grown in the ampoules on a small amount of suitable agar medium. If the culture does not survive using glycerol as the cryoprotectant there are several alternatives that can be used [such as 10% dimethyl sulphoxide (DMSO) or a mixture of 5% DMSO and 8% glucose (Smith, 1983a)]. The optimum cooling rate should always be used.

VI.   STORAGE AND SURVIVAL BY SUBCULTURE

A. On agar at growth temperatures

The storage conditions of growing cultures are often dependent on the physiological requirements of each fungus. In addition to providing a suitable growth medium, temperature, light and humidity must also be taken into account. The growth of cultures at room temperature is perhaps the most convenient method of maintenance. Keeping culture vessels in racks and storing them on shelves between transfers is suitable for the majority of microfungi, which normally grow well at temperatures of 20–22°C. However, some fungi are more sensitive to temperature and it is necessary to establish the range in which an isolate will grow before selecting a suitable storage temperature. Chytrids and other water moulds grow better at temperatures lower than 20°C, whereas some thermophiles require temperatures between 30–50°C for growth.

The light requirement of a fungus for spore formation will influence the choice of storage condition. Some fungi, such as *Pyronema domesticum*, require daylight to induce production of fruiting bodies and sporulation. Others, such as some species of *Coprinus*, respond better to growth in the dark to produce sporophores. Many dematiaceous hyphomycetes, coelomycetes, and ascomycetes require the stimulation of near ultraviolet or black light before they will produce spores (Leach, 1971).

Most fungi grow well at a fairly high humidity (Onions, 1971), and limiting water availability will certainly hinder their growth. Normally, within culture vessels, the moisture in the atomsphere above the culture provides an adequate climate.

A collection of growing fungi must be checked at regular intervals to ensure that cultures have not become desiccated. The rate at which the cultures lose moisture will depend upon the conditions under which they are stored. High temperatures and low humidities will inevitably mean that transfers should be more frequent. The Centraalbureau voor Schimmelcultures (CBS), Netherlands, recommends temperatures of 16–17°C and a relative humidity of 60–70% for storage (von Arx and Schipper, 1978). Keeping cultures in high humidities will allow condensation and growth on the outside of culture vessels and will lead to cross-contamination.

Transfers to fresh media are normally made every 3–6 months. Some cultures, such as the water moulds and human and animal pathogens (Onions, 1971), require more frequent transfer, and collections kept in the tropics or low humidity climates require even more frequent attention (Fennell, 1960). The genera *Allomyces, Achlya, Isoachlya, Phytophthora, Pythium* and *Saprolegnia* and the basidiomycete genera *Boletus, Coprinus, Corticium, Cortinarius* and *Mycena* require subculturing either once a month or every 2–3 months (von Arx and Schipper, 1978).

B. On agar in the refrigerator

The use of cold storage to decrease the rate of metabolism, which increases the period between transfers to fresh media, has proved very successful. Normal domestic or laboratory refrigerators, which give temperature ranges of 4–7°C, may be used. The cultures are stored in racks or boxes on the shelves to give easy access. Overpacking of refrigerators that are in regular use can cause problems due to build-up of condensation and the probability of cross-contamination (Dade, 1960). A yearly transfer is usually adopted for these cultures, though species which fail to survive this storage period must be transferred more frequently. Most moulds and common fungi, such as *Penicillium, Aspergillus* and the Mucorales, survive well at 5–8°C but some, such as *Piptocephalis*, are sensitive to cold (Onions, 1971). A decline in the antibiotic titre of *Penicillium chrysogenum* (MacDonald, 1972;

Wellman, 1971) has been observed with storage at 4°C though the growth and sporulation of this species seems to be unimpaired at this temperature.

At the International Mycological Institute (IMI) all cultures are kept at 4–7°C, pending the results of storage by other means. Storage periods are normally between 4 and 6 months, but many cultures have survived for 1–2 years when left in the refrigerator. Although specimens have been shown to dry out after 9 months, this depends on the medium on which the fungus is growing. In general, the Zygomycotina, Ascomycotina, Basidiomycotina and Deuteromycotina successfully survive storage at 4–7°C for periods up to 1 year, whereas the Mastigo-mycotina survive less well.

C. On agar under oil

Using the method reported by Buell and Weston (1947) the entire collection at IMI was stored under oil following trial experiments in 1950 (Table I). Since 1956 the number of isolates stored by this method has increased to over 14 000. Although many have survived storage for over 30 years, a number have deteriorated or died. Forty genera require subculturing every 2 years because they have shown deterioration during storage (Table II). Some

Table I. Survival of some cultures from the IMI oil collection

	IMI number	Survival (years)
Aspergillus avenaceus	16140	32
A. citrisporus	25285	32*
A. medius	29188 ii	20
A. restrictus	39044	24
Botryosphaeria obtusa	38560	32
B. ribis	36476	32
Ceratocystis paradoxa	37270	32
C. paradoxa	39075	32
C. radicicola	36479	32*
Chlamydomyces palmarum	39639	32
Corticium rolfsii	33912	32
C. rolfsii	37955	25*
Eleutherascus terrestris	25845	32*
Helicodendron triglitziense	38968	32*
Helminthosporium (Drechslera) portulacae	37710	32*
Humicola sp.	38777	32*
Mortierella alpina	38598	12 D
Mucor circinelloides	39478	23

Table I. *Continued.*

	IMI number	Survival (years)
Nectria pityrodes var. *saccharina*	37228 a	32*
Penicillium adametzii	39751	25*
P. thomii	39740	32
P. cyaneum (Bainier & Sartory) Biourge	39744	25
P. klebahnii	39737	32*
P. fellutanum	39734	32
P. velutinum	39747	32
P. javanicum	39733	32*
P. lapidosum	39743	32
P. lineatum	39741	32
P. rasile	39735	32
P. lividum	39736	32
P. oxalicum	39750	32
P. sclerotiorum	39742	32
P. shearii	39739	32
P. spinulosum	39749	32
P. turbatum	39738	32
P. waksmanii	39746	25 (sectoring)
Petriella sordida	38601	32
Phytophthora citricola	21173	32
P. nicotianae	22176	32
Podospora fimbriata	38111 i	10D
P. fimbriata	38111 1ii	32
P. fimbriata	38111 2i	32
P. fimbriata	38111 2ii	32
Rhizoctonia oryzae-sativae	31287	32
Sclerotium coffeicola	39753	32*
Scopulariopsis sp.	16401	32*
Scopulariopsis sp.	16404	32*
Setosphaeria rostrata	22971	32
Sporendonema casei	37084	17D
Stephanosporium cereale	38105	32
Thanatephorus cucumeris	20697	32D
Thanatephorus praticola	37886	32
Thielaviopsis basicola	36482	12
Torula herbarum	31291	32
T. ligniperda	36123	32
Ustilago scitaminea	36859	32
Verticillium theobromae	31432 a	32
Volutella ciliata	38780	32

* Showed inconsistency on retrieval and required several isolations to
  achieve good growth.
D Died between number of years survival indicated and the next viability test.

Table II. Genera and number of isolates stored under oil in the IMI Culture Collection that require regular transfer every 2 years because of deterioration during storage

Genus	No. of isolates	Genus	No. of isolates
Acremoniella	1	Helicosporina	1
Alternaria	4	Heliscus	1
Arthrobotrys	14	Hypomyces	3
Basidiobolus	7	Khuskia	1
Beltrania	4	Lophiostroma	2
Beltraniella	1	Melanospora	2
Calonectria	3	Mesobotrys	1
Cercospora	16	Monacrosporium	4
Chaetospnaeria	5	Monascus	4
Chalaropsis	1	Mortierella	58
Chrysosporium	1	Mycovellosiella	5
Claviceps	3	Nectria	5
Colletotrichum	81	Nodulisporium	41
Cylindrocarpon	3	Penicillifer	2
Cytospora	1	Periconia	24
Echinosporangium	1	Pseudocercospora	1
Endophragmia	1	Pyrenochaeta	5
Georgefischeria	1	Pyrenophora	24
Gloeosporium	4	Spermospora	1
Glomerella	8	Stigmina	1

fungi, such as strains of *Fusarium*, some Saprolegniaceae and other water moulds, are not stable or do not survive long storage periods under oil and so require subculturing at 3–6 month intervals. However, it has been reported by others (Reischer, 1949) that Saprolegniaceae and other water moulds survive 12–30 months. As with most storage techniques, survival seems to vary for individual strains within a species. The frequency of transfer is also affected by the poor condition of the isolate or failure to sporulate in culture.

Although storage under mineral oil has the advantage of being inexpensive and simple, it has the major disadvantage of allowing growth under what may be adverse conditions. The slower growth rate induced may allow naturally occurring mutants to become dominant over the wild-type. The occurrence of sectoring as an expression of mutation is increased after storage under a layer of mineral oil. Constant supervision is required by a specialist to ensure that the original strain, rather than the variant or mutant, is transferred.

D. On agar blocks in water

This simple method of storage has been used for the maintenance of fungi for many years. Fungi pathogenic to humans (Castellani, 1939; Castellani, 1967), plant pathogenic fungi (Figueiredo, 1967; Figueiredo and Pimentel, 1975; Boeswinkel, 1976), oomycetes (Clark and Dick, 1974), entomophthorales, pyreno-mycetes, hymenomycetes, gasteromycetes, hyphomycetes (Ellis, 1979) and ectomycorrhizal fungi (Marx and Daniel, 1976) have all survived using the technique.

At IMI the method has been used successfully with the genera *Phytophthora* and *Pythium*, both of which have survived 2 years' storage without loss in viability. Many of the strains of *Phytophthora* remained viable for 5 years or more. However, their ability to infect their hosts diminished after 2–3 years storage.

VII. STORAGE AND SURVIVAL OF DRIED CULTURES

A. On anhydrous silica gel

Many fungi have been successfully stored by this technique (Onions, 1977; Smith and Onions, 1983b), but it is not satisfactory for mycelial cultures or for species belonging to the Mastigomycotina. It has proved to be a good technique for maintaining cultures in a stable condition, since many strains of genetic importance have been preserved without change (Ogata, 1962; Perkins, 1962; Barratt *et al.*, 1965). The pathogenicity of *Helminthosporium* (*Drechslera*) *maydis* was maintained for 1 year using this method (Sleesman *et al.*, 1974).

The storage of 421 isolates was examined at IMI and the results, broken down into major taxonomic groups according to Hawksworth *et al.* (1983) and the *Index of Fungi* (Anon, 1971–1981), are shown in Table III. Although 306 (73%) survived for periods up to 11 years, 115 failed to survive (Smith and Onions, 1983b). It appeared that survival was isolate-specific. The 115 failures have survived other storage techniques, and a list of these isolates, with their survival period, is given in Table IV. The limiting periods for survival are not yet known because storage continues. At present, 50% of strains have survived for 18 years.

Table III. Viability of all the isolates processed and stored on silica gel at IMI following 8–18 years storage

Genus	No. of genera viable	No. of species viable	No. of isolates tested	Viable	% Viability
Myxomycota					
Acrasiomycetes	0	0	1	0	0
Eumycota					
Mastigomycotina					
Chytridiomycetes	0	0	5	0	0
Oomycetes	0	0	5	0	0
Zygomycotina					
Zygomycetes	9	17	34	20	59
Ascomycotina					
Hemiascomycetes	3	4	5	5	100
Plectomycetes	5	9	13	11	85
Pyrenomycetes	17	23	54	35	63
Discomycetes	2	2	6	2	33
Basidiomycotina					
Hymenomycetes	8	10	19	10	52
Gasteromycetes	0	0	1	0	0
Deuteromycotina					
Hyphomycetes	38	136	250	202	81
Coelomycetes	7	18	28	21	75

Viability of all isolates tested 1971–82:

Tested	Failed to survive during storage	Viable	% Viability
421	115	306	73

Table IV: Survival (years), using various methods, of the 115 fungi that failed to survive or died in storage in silica gel†

	IMI number	Method			
		Silica gel	Oil	Freeze-drying	Liquid N$_2$
Allomyces arbuscula	129543	—	14	—	7
A. arbuscula	152201	5	7D	0	0
*A. javanicus**	144364	—	4D	0	0
Armillaria mellea	61755	—	27	0	0
Arnium arizonense	169785	—	8	8	8
Arthrobotrys oligosporus	102121	<3	16	6	0
Ascochyta fabae	135517	—	7P	12P	0
Ascotricha lusitanica	147693	—	11	11	11

Table IV: *Continued.*

	IMI number	Method			
		Silica gel	Oil	Freeze-drying	Liquid N_2
Aspergillus candidus	130667	<3	14	8	0
A. citrisporus	25285	4	11P	13	0
A. ochraceus	16247iii	2	11	11P	0
A. restrictus	127782	<3	14	14	0
A. ustus	100391	3	18	8	0
A. wentii	162039	4	11P	1	0
Basidiobolus meristosporus	108476	<3	12	5	0
Beltraniella sp.	223748	—	4	2	0
Botryodiplodia theobromae	125847	4	9	8	0
Calonectria quinqueseptata	136139	<3	11	—	11
Candida lipolytica	93743	—	20	15	0
Ceratocystis ulmi	147188	2	9P	1	0
C. ulmi	173135	2	9P	1	0
C. ulmi	173136	3	9	9	0
*Cercospora beticola**	77043	—	4P	0	0
C. sesami	111779	—	12P	12ST	12
Chaetomidium fimeti	116692	—	16	10	0
*Chytridium olla**	86666	—	7D	0	—
Cochliobolus sativus	166172	—	9	9	9
C. sativus	166173	—	9	9	9
Coemansia formosensis	170166	—	9	9	8
*C. mojavensis**	140079	1	0	0	0
C. pectinata	142377	2	10P	9	10
Colletotrichum gossypii	82269	2	5ST	12	12
C. trichellum	82378	<4	15	10	0
C. truncatum	86431	1	21	5	0
Conidiobolus coronatus	68174	2	13	—	0
C. heterosporus	102043	—	18	0	0
C. lobatus	138635	—	12	—	0
C. mycophagus	113701	<3	6	—	9
*Coprinus luteocephalus**	161421	—	0	9	0
*C. viarum**	161423	—	0	9	0
Corticium rolfsii	77445a	—	22	5	0
Culicinomyces clavosporus	177011	<6	6	6D	0
Dictyostelium discoideum	69094ii	—	27	15	11
Exobasidiellum culmigenum	136517	—	13	7D	0
Fusarium culmorum	175485	<7	9	9	0
F. solani	76761	1	21D	15	0
Gaeumannomyces graminis	160145	—	11	2	2
Gelasinospora sp	47702	—	24	13	0
Geosmithia lavendula	40570	1	27	15	0
Gnomonia comari	164147a	—	9	9	0

Table IV: *Continued.*

	IMI number	Method			
		Silica gel	Oil	Freeze-drying	Liquid N₂
Helicodendron tubulosum var. *phialosporum*	92743	<3	12	6	10
Helicosporina veronae	114458	—	16	4	8
Heliscus submersus	82609	<3	16	9	9
Heterocephalum aurantiacum	131684	8	12P	12	14
Hypoxylon mediterraneum	75991	<3	22	0	0
H. nummularium	146051	0.2	6	—	0
Leptosphaeria doliolum subsp. *pinguicola*	199777	<4	6	0	5
Martensiomyces pterosporus	60573	—	26	2	2
Metarhizium anisopliae	98375	<1	13P	10	0
Micromonospora vulgaris	126892	5	15P	11	0
Monacrosporium salinum	109555	—	17	5	5
Mortierella bainieri	167609	—	9	9	0
Mycosphaerella deightonii	119431	—	7	—	0
Mycovellosiella ferruginea	124973	—	15	9	0
Neurospora crassa	19419	5	12	6	0
N. crassa	68614ii	—	12	12	12
N. crassa	147001	2	5D	0	0
Normuraea atypicola	186963	<1	8	7	0
Ophiovalsa suffusa	173497	<3	9	9	9
Penicillium aurantiogriseum	19759	<5	19	15	0
P. brevicompactum	17456	<3	25	15	0
P. canescens	149218	—	11	11	0
P. corylophilum	101082	2	19	11P	0
P. digitatum	91956	—	14	14	0
P. digitatum	92217	<3	12	14	0
P. expansum	191204	2P	7	7	0
P. hirsutum	68414	<3	22	12	0
P. janthinellum	108033	<4	18	15	0
P. luteum	112513	—	5D	11	0
P. phoeniceum	148393	—	12	12	12
P. spirillum	197479	2P	7	6	0
Phycomyces blakesleeanus	118496	5	12P	8	0
P. blakesleeanus	118497	11	11	12	0
Phoma epicoccina	164070	—	9	9	9
Phytophthora cactorum	21168	—	32	—	10
Piptocephalis xenophila	156650	4	10P	5P	11
Pleospora infectoria	173200	3	9	9	9
Pyrenopeziza brassicae	204290	—	0	3	0
Pyrenophora graminea	129760	<1	15ST	10D	0
Pyronema domesticum	57472	—	27	15	2

Table IV: *Continued.*

	IMI number	Method			
		Silica gel	Oil	Freeze-drying	Liquid N$_2$
Pythium debaryanum	48558	—	28	0	0
P. flevoense	176046	—	8	0	7
P. middletonii	42098	—	31	0	10
Rhizoctonia carotae	162910	—	10	0	0
Rhizophydium sphaerotheca	143633	—	12	—	—
Rhizopus rhizopodiformis	158738	—	11	11	0
Rhodotorula rubra	38784	4	12	5	0
Ryparobius polysporus	75299	—	13	10	0
Saprolegnia ferax	146489	—	11	—	3
Sclerotinia sclerotiorum	147201	—	12	0	12
Sclerotium delphinii	159926	—	10	4	9
Seiridium sp.	151978	8	11	10P	0
Serpula lacrimans	152233	—	11	0	0
Sphaerobolus stellatus	155101	—	10	0	10
Sporobolomyces roseus	43529	—	31	11	0
Stachybotryna columnare	158980	—	10	0	0
Stilbum macrosporum	163252	—	9	9	0
Streptomyces griseus	50967	—	29	15	0
Streptomyces lisandri	137178	—	12	10	0
Syzygites megalocarpus	231978	<2	0	2	4
Thielavia terricola	153731	—	1D	11	11
Trichoderma viride	57421	5	25	4	0
Trichothecium roseum	129425	—	14	14	0
Tritirachium oryzae	169856	<3	10	10	0
Zalerion maritima	81620	<3	14	—	10
Zopfiella leucotricha	153733	—	1	10	10

D — dead at time of examination.
P — poor revival at time of examination.
ST — sterile at time of revival.
* These isolates were discarded from the collection after the initial processing as they were atypical.
† More recent survival figures on some strains of the listed species are available from the International Mycological Institute, Kew, Surrey, UK.

B. In soil/sand

Fusarium, a genus of fungi renowned for its instability in culture, has been kept in a stable condition by storage in soil (Gordon, 1952; Booth, 1971b). *Septoria* species have been kept by this method without loss of sporulation or pathogenicity (Shearer *et al.*, 1974), and storage of *Pseudocercosporella herpotrichoides* has been very successful (Reinecke and Fokkema, 1979).

At IMI the technique has been used successfully for over 20 years to store 764 isolates of *Fusarium* and other genera (Booth and Butterfil, unpubl. data) (Table V). Not all of the isolates have been tested recently, but of 54 examined after 10–20 years' storage, 48 have good growth and six have died. Of the isolates that have died in storage all had survived for more than 10 years. It is felt that the considerable time-lag before the onset of dormancy due to dryness might be sufficient for mutant vegetative strains to overgrow the wild-type. An examination of a collection developed by Gordon (1952) showed that 75% of *Fusarium equiseti*, 75% of *F. semitectum* and 50% of *F. acuminatum* isolates had been outgrown by mutant strains (Booth, 1971b).

Table V: Isolates in the IMI collection that have been stored in soil

Genus	No. of species	No. of isolates
Calonectria	6	8
Cylindrocarpon	12	32
Cylindrocladium	5	9
Fusarium	55	652
Gibberella	4	6
Melanospora	2	9
Nectria	17	47
Thielavia	1	1

VIII. STORAGE AND SURVIVAL OF FREEZE-DRIED CULTURES

Although many fungi have been preserved by freeze-drying (Smith, 1983b) it is a technique that seems to be suitable only for sporulating isolates. Mycorrhizal fungi generally give poor results (Jackson *et al.*, 1973; Crush and Pattison, 1975), although with a modified technique cultures that are normally difficult to preserve can survive. Pre-freezing to −45°C of plant tissue infected with *Pythium acanthicum* and *P. irregulare*, followed by drying under vacuum at room temperature, facilitated the preservation of these pathogens (Staffeldt and Sharp, 1954). *Puccinia* urediniospores, pre-frozen to between −45°C and −50°C and dried under vacuum at −10°C, survived freeze-drying (Sharp and Smith, 1952).

At IMI over 11 000 isolates have been freeze-dried, of which over 10 000 have survived. A list of genera and numbers of isolates that have failed to survive centrifugal freeze-drying is given in Table VI. As with most preservation techniques, success with freeze-drying varies between isolates of the same species, although preservation is more likely to be successful if good healthy cultures are used. A list of genera that have survived 20 years storage (Table VII) and those that have survived the technique without suspending medium on agar blocks (Table VII) indicates the success of the technique.

Table VI: Genera that have failed to survive centrifugal freeze-drying

Achlya	Herpotrichia	Pythium
Allomyces	Kretschmaria	Quarternaria
Areolospora	Lacellinopsis	Saprolegnia
Armillaria	Lasiobolidium	Searchomyces
Arthrocladium	Lentinus	Selenosporella
Ascocalvatia	Lenzites	Selinia
Balansia	Leptoporus	Sigmoidea
Batteraea	Lomachashaka	Sphaerobolus
Biscogniauxia	Marasmius	Spondylocladiopsis
Calospora	Monotosporella	Stereum
Camposporium	Nummularia	Sympodiella
Chytridium	Panus	Syzygites
Cladobotryum	Penicillifer	Tetracladium
Coriolus	Phaeoisariopsis	Tetranacrium
Dactuliophora	Phyllosticta	Unbelopsis
Eleuthersacus	Physarum	Urohendersonia
Entomopthora	Phytophthora	Ustilaginoidea
Eremomyces	Piedraia	Ustulina
Fomes	Platystomum	Volvariella
Ganoderma	Puccinia	

IX. STORAGE AND SURVIVAL OF FROZEN CULTURES

A. On agar slopes at −20°C

Many cultures frozen on agar slopes and stored at −20°C can be expected to survive 4–5 years if they are not allowed to thaw. Some fungi are sensitive to this form of storage and die when frozen to this temperature. *Martensiomyces*, some oomycetes and water moulds fall into this group. The IMI taxonomic collection

Table VII. Genera that have survived 14 years since centrifugally freeze-drying at IMI.*

Absidia	Elsinoe	Phialocephala
Acremonium	Eremothecium	Phialophora
Actinomucor	Fusarium	Phoma
Alternaria	Gelasinospora	Phomopsis
Amorphotheca	Geomyces	Phycomyces
Arthroderma	Geosmithia	Pirella
Ascotricha	Geotrichum	Pithomyces
Aspergillus	Gibertella	Pleurophragmium
Aureobasidium	Gliocladium	Rhinotrichum
Bispora	Gliomastix	Rhizopus
Botryotrichum	Glomerella	Saccharomyces
Byssochlamys	Hirsutella	Scopulariopsis
Cephaliophora	Humicola	Sesquicillium
Cephalosporium	Leptographium	Setosphaeria
Cercospora	Loramyces	Spegazzinia
Chaetodiplodia	Mammaria	Sporobolomyces
Chaetomidium	Mariannaea	Sporophora
Chaetomium	Memmoniella	Sporothrix
Chaetompsina	Microascus	Stachybotrys
Chalara	Monodictys	Staphylotrichum
Chloridium	Mucor	Stemphylium
Circinella	Mycosphaerella	Stilbella
Cladosporium	Myrothecium	Streptomyces
Coniella	Nectria	Sydowia
Coniothyrium	Neocosmospora	Syncephalastrum
Cordyceps	Neodeightonia	Thamnostylum
Coryne	Neurospora	Thermoascus
Cunninghamella	Oidiodendron	Thielavia
Curvularia	Ophiobolus	Trichoderma
Cylindrocarpon	Paecilomyces	Trichophaea
Didymella	Penicillium	Ulocladium
Didymosphaeria	Pestalotiopsis	Valsa
Doratomyces	Petriellidium	Venturia
Drechslera	Phaeotrichoconis	Volutella
Eladia	Phialomyces	Zygorhynchus

* More detailed information is available from IMI (IMI Culture Collection database)

of *Aspergillus*, *Penicillium* and related genera were kept by this means and have survived for 5 years.

Cold storage of fungi has been used for medically important fungi (Carmichael, 1956, 1962) and proves to be even more successful if the stock cultures are not allowed to thaw during the retrieval procedure (Kramer and Mix, 1957). Freezing and

Table VIII: The survival periods of isolates freeze-dried without suspending medium on agar blocks

	IMI number	Survival (years)
Acremonium sp.	55286	14
Ascocoryne sarcoides	68130	14
Aspergillus amstelodamii	71295	8
A. candidus	73074	14
A. carneus	73777	14
A. nidulans var. echinulautus	61454ii	14
A. niger	75353ii	14
A. quadrilieatus	72733	14
A. ostianus	93445	14
Chaetomium nigricolor	114513	14
Curvularia trifolii f. sp. *gladioli*	75377	13
Cylindrocarpn congoense	69504	14
Embellisia chlamydospora	67737	14
Fusarium graminearum	69695	14
Nectria gliocladioides	71095	14
Paecilomuces dactylethromorphus	65752	14
Penicillium citrinum	72029	14
P. echinulatum	68236	14
P. janthinellum	71625	13+
P. nigricans	96660	14
P. roquefortii	129207	14
P. simplicissimum	68220	14
P. spinulosum	68617	14
Pestalotiopsis gracilis	69749	14
Phaeotrichoconis crotalariae	69755	14
Phialomyces macrosporus	110130	14
Phomopsis oncostoma	68344	14
Pycnoporus sanguineus	75002	9
Sagenomella griseoviridis	113160	13+
Scopulariopsis carbonaria	86941	14
Sporidesmium flexum	246524	1

thawing microbes has many associated problems (Calcott, 1978), and it has been found that some fungi are susceptible to freezing damage without cryoprotectants. However, storage at −20°C in 10% (v/v) glycerol gives poorer survival in many cases than freezing on an agar slope without glycerol. This is probably a result of the fungus cell not freezing at this temperature in the presence of glycerol (Morris *et al.*, 1988), allowing biochemical reactions to continue and hypertonic solution effects to take place.

B. In liquid nitrogen

Freezing and storing fungi in liquid nitrogen, or the vapour phase above it, has proved very successful. Of the 5500 isolates preserved in this way at the IMI, 5114 have survived up to 22 years. The Mastigomycotina proved to be the most difficult to preserve by this method, although the isolates that grew well in culture survived well (see Table IV). This group of fungi seem to suffer from mechanical damage incurred during the preparation of suspensions. If precautions are taken not to cause damage of this kind, many of this group survive well (Smith, 1982). Pre-growth of cultures in the refrigerator at 4–7°C can improve the viability of some fungi that are normally difficult to freeze.

It is the selection of the optimum cooling rate for the individual fungal strain that has the greatest effect on survival at temperatures of liquid nitrogen or its vapour. A study sponsored by the Commission of the European Communities (CEC) Biotechnology Action Programme (BAP) has shown that many fungi respond differently to cooling and have optimum cooling rates for recovery (Tables IX and X). It is essential that optimum cooling conditions are determined and used prior to storage at low temperatures, to ensure long-term survival and stability of the fungus properties.

The BAP study also showed that storage temperature dramatically influenced the viability of the preserved strains. Of those stored at −20°C, 14 strains (14%) failed to recover following 1 year of storage and a further 12% showed a 90% loss in viability during the same period. Ten strains (10%) failed to recover from −40°C storage, and a further 7% showed a 90% fall in viability. In contrast, only two strains failed to survive liquid nitrogen storage (−196°C) (*Phytophthora humicola* and *Volvariella volvacea*), and survival levels of other strains remained high.

Storage in liquid nitrogen has proved successful for a wide range of fungi (Hwang, 1966, 1968; Hwang *et al.*, 1976; Butterfield *et al.*, 1974; Alexander *et al.*, 1980). Fungi that are difficult or impossible to grow in culture can be kept alive for long periods in liquid nitrogen. Examples are rust and smut spores (Loegering, 1965; Kilpatrick *et al.*, 1971) and *Sclerospora* species (Gale *et al.*, 1975; Long *et al.*, 1978; Smith, 1982).

Many storage containers other than glass ampoules can be used successfully in liquid nitrogen vapour. When polypropylene ampoules were used, better revival of *Basidiobolus*, *Rhizophydium*, *Phytophthora* and *Serpula* was achieved than with glass ampoules (Butterfield *et al.*, 1978). Although plastic ampoules with screwcaps may leak during storage, they are a safer alternative to glass because they do not explode on expansion of liquid nitrogen

Table IX. Quantitative cryomicroscopy of fungal hyphae (Morris *et al.*, 1988)

Species	Critical cooling rate (°C min^{-1})*	Nucleation temperature (°C)
Mastigomycotina		
Achlya ambisexualis	6	−6 to −18
Phytophthora humicola	53	−8 to −17
P. nicotianae	>120	No ice seen
Pythium aphanidermatum	16	−6 to −14
Saprolegnia parasitica	4	−2 to −7
Zygomycotina		
Mortierella elongata†	18.5	−9
Mucor racemosus	10	−11 to −14
Ascomycotina		
Sordaria fimicola	4	−8.5
Basidiomycotina		
Lentinus edodes	>100	No ice seen
Schizophyllum commune	9	−25 to −30
Serpula lacrymans		
(i) hyphae from 7–21-day cultures	15.5	−10
(ii) hyphae from 28-day cultures		
(a) <4.45 μm in diameter	>100	No ice seen
(b) >4.45 μm in diameter	1	−15.5
Sporobolomyces roseus	>100	No ice seen
Volvariella volvacea	>100	No ice seen
Deuteromycotina		
Alternaria alternata	12.5	−14.5
Aschersonia alleyrodis	>100	No ice seen
Aureobasidium sp.	>100	No ice seen
Penicillium expansum	18	−14
Trichoderma viride	5	−5 to −11
Trichophyton rubrum	18	−18
Wallemia sebi	9	−18

* Rate of cooling when 50% of cells observed contain ice.
† The results presented here are those after growth in liquid medium.

(Simione *et al.*, 1977). Leakage may be prevented by the use of polypropylene sleeves (supplied by Nunc Cryoflex, Gibco Life Technologies Ltd). The use of plastic drinking straws that can be autoclaved are a space-saving alternative (Dietz, 1975; Elliott, 1976; Stalpers *et al.*, 1987), although storage conditions must not allow contact with liquid nitrogen, which may penetrate and cause contamination or splitting of straws on warming. Polyester film packets containing fungal material can also be used (Tuite, 1968).

Table X. The optimum cooling rate and recovery of 20 species of fungi suspended in either growth medium or glycerol (10% v/v) (Morris *et al.*, 1988)

Species	Growth medium		Glycerol	
	Optimum cooling rate ($°C min^{-1}$)	Recovery (%)*	Optimum cooling rate ($°C min^{-1}$)	Recovery (%)
Mastigomycotina				
Achlya ambisexualis	—	0	9	88
Phytophthora humicola	—	0	3	69
P. nicotianae	—	0	0.5–11.0	72–91
Pythium aphanidermatum	—	0	8–29	52–53
Saprolegnia parasitica	—	0	10	32.5
Zygomycotina				
Mortierella elongata	10–200	95	1–200	100
Mucor racemosus	25	36	5	31
Ascomycotina				
Sordaria fimicola	1–200	*c.* 100	1–200	*c.* 100
Basidiomycotina				
Lentinus edodes	1	27	1.0–3.5	100
Schizophyllum commune	1	92	23	92
Serpula lacrymans				
(i) hyphae from 7–21-day cultures	1	0	0.5	99
(ii) hyphae from 28-day cultures				
(a) <4.45 µm in diameter	—	0	13	20
(b) >4.45 µm in diameter	—	0	13	20
Sporobolomyces roseus	4.5	*c.* 70	1	100
Volvariella volvacea	—	0	1	38
Deuteromycotina				
Alternaria alternata	21.5	88	20	97
Aschersonia alleyrodis	10.5	86	1	100
Aureobasidium sp.	10.2	83	200	91
Penicillium expansum	0.5–200	85–100	1–12	75–90
Trichoderma viride	3.8	*c.* 100	3.8–10.0	*c.* 100
Trichophyton rubrum	4	70.5	9	86
Wallemia sebi	83	*c.* 100	77	*c.* 100

* For recovery method see Morris *et al.* (1988).

X. BLACK LIGHT: ITS USE TO INDUCE SPORULATION IN FUNGAL CULTURES

Successful fungal preservation sometimes depends on the presence of spores. Light is very important in this respect, and short wavelengths have been used to induce spore production (Leach, 1962, 1971). Near ultraviolet light or black light of wavelength 310 to 400 nm may affect pigmentation, the gross morphology of the colony or even spore morphology, though the effects are not sufficient to interfere with identification.

The light-benches at IMI have three 1200 mm fluorescent lamp holders 130 mm apart. A black-light tube (Philips TL 40 W/08) is held in the centre holder and a cool-white tube (Philips MCFE 40 W/33) is placed on either side. They are controlled by a time switch which is set to a 12 h on/off cycle. The fungi in disposable plastic Petri dishes are supported on a shelf 320 mm below the light source. The fungi are inoculated onto the plate and allowed to grow for 2–4 days before placing under the light. The edges of the dishes are sealed with transparent adhesive to prevent rapid drying (Booth, 1971b).

The growth medium can have an effect on stimulation and weak media should normally be used to encourage sporulation (Leach, 1962). When growing on glucose casein hydrolysate medium and stimulated by near ultraviolet light, *Diaporthe phaseolorum* var. *batatis* produces many perithecia and ascospores. However, when grown on malt or potato glucose media, the number of perithecia is much lower (Timnick *et al.*, 1951).

XI.   MITE PREVENTION

Mites, commonly of the genera *Tyrotlyphus* and *Tarsonemus*, can cause damage to fungal cultures in two ways. First, they can eat the culture and completely destroy it. Second, they may carry fungal spores and bacteria on their bodies and, as they move from one culture to another, cross-contamination may occur. Methods of control used by different workers are varied and a combination of precautions seems to be most effective. Hygiene, fumigation, mechanical and chemical barriers, and protected storage are the normal methods of prevention (Anon, 1982).

A. Hygiene

Hygiene coupled with a quarantine procedure is perhaps the best protection. All work surfaces should be kept clean and cultures should be protected from aerial infection by storage in cabinets. Benches should be washed down with an acaricide. Acteltic (ICI Agrochemicals) is suitable. However, such chemicals are hazardous and extreme caution must be exercised in their use. A 'dirty' room should be available, to which all incoming cultures are directed before ensuring that they are mite-free. When mites are found, the infested cultures should be sterilized and fresh isolates sought. If this is impossible, the culture should be covered with a quantity of liquid paraffin or stored in a −20°C freezer for 2–3 days and subcultured at a later date.

B. Fumigation

Unpurified tractor vapourizing oil was formerly used at IMI and, although not an acaricide, it deterred mites. *Para*-dichlorobenzene is effective, although it can produce abnormal growth in fungi; camphor is another fumigant that can be used successfully (Smith, 1967). Fumigants are placed in the storage cabinets either as a short treatment or for permanent storage, provided that the chemicals used do not affect the fungi and are not harmful to personnel.

C. Mechanical and chemical barriers

Surrounding cultures with oil, water or vaseline can prevent infection from crawling mites, although those carried on clothing, by insects or by air currents can still cause infestation.

Sealing the necks of tubes or bottles with sterile cigarette paper attached with copper sulphate gelatine glue (20 g gelatine, 2 g copper sulphate in 100 ml distilled water) can prevent mite infection, while still allowing air to pass and so not impeding fungal growth (Snyder and Hansen, 1946). Smith (1971) recommended the use of disposable plastic bottles with plastic caps that can be screwed down tightly to prevent the entry of mites.

D. Cold storage

Although cold storage at 4–8°C reduces the spread of mites, they soon become active on removal from the refrigerator. Deep-freezing below −18°C will kill both the mite and most of its eggs.

6. MAINTENANCE OF YEASTS

B. E. KIRSOP

Microbial Strain Data Network, Institute of Biotechnology,
Cambridge University, Cambridge, UK

I. Introduction 161
II. Maintenance methods available 163
III. Subculturing 163
   A. Subculturing in broth......................... 164
   B. Subculturing on agar slants 165
   C. Subculturing on agar slants with oil overlay 166
   D. Subculturing strains with special properties 167
   E. Subculturing in water 168
IV. Drying 168
   A. Paper replica method 168
   B. Silica gel method 169
V. Freeze-drying 171
   A. Centrifugal freeze-drying 171
VI. Freezing 175
   A. Freezing in liquid nitrogen (ampoules) 175
   B. Freezing in liquid nitrogen (straws) 177
   C. Stability of genetically modified yeasts under
      different cryopreservation regimes 180

I. INTRODUCTION

In general, yeast cells are considered to be robust, tolerant of unfavourable conditions, nutritionally undemanding, and readily managed in industry. It is assumed, therefore, that they are also easy to maintain. However, a number of different maintenance methods commonly used result in poor survival levels and instability of properties.

The factors affecting survival are becoming better understood at the subcellular level, but many strains from a wide range of species remain difficult to preserve. The relatively poor performance of yeasts may be attributed in part to the large size of cells compared with bacteria and the absence of the kind of resistant spores produced by many of the higher fungi. In the light of present knowledge, high survival rates can best be achieved by

MAINTENANCE OF MICROORGANISMS 2nd Edn
ISBN 012–410351–0

careful attention to growth conditions, choice of cryoprotectant, cooling rate and careful attention to the techniques used for preservation and revival.

Percentage survival of the total population levels following subculturing and drying are generally low although 'cultures' may appear viable (Kirsop, 1974, 1978). Survival following freeze-drying is also frequently poor, but may be improved in some strains by careful selection of suspending medium (Berny and Hennebert, 1989). In contrast, survival following storage in liquid nitrogen is high, often reaching levels of 100% (Hubalek and Kochova-Kratochvilova, 1978). There appears to be no relationship between survival and taxonomic position, and the factors determining survival are specific for each strain. This means that a maintenance method that is satisfactory for one strain may be unsuitable for others, and generalizations on the effects of maintenance method on survival should be viewed with caution, particularly if studies have been conducted on one or two strains only. If only a few yeasts are to be maintained, it may be possible to establish the optimum conditions for each strain; if larger numbers are involved, it may only be practicable to select the best method for most of the strains.

If strain stability is of paramount importance the choice of maintenance method becomes critical. Any method that enables cell division to occur should be rejected. It has been shown (Kirsop, 1974) that following freeze-drying yeasts in general remain unchanged morphologically, physiologically and with regard to industrial properties, although other authors have shown that genetic change can occur (Souzu, 1973). Again, some laboratories have found that yeasts dried on silica gel may show substantial changes (Bassel et al., 1977; Kirsop, 1978), whereas others report satisfactory results (Woods, 1976). It has been reported by Bassel et al. (1977) that genetically marked strains retain their characteristics after drying on filter paper, and there is growing evidence that many strains stored in liquid nitrogen remain stable (Wellman and Stewart, 1973; Hubalek and Kochova-Kratochvilova, 1978; Pearson et al., in press). Factors affecting the stability of genetically manipulated strains have been studied and a section (VI.C) on this aspect of yeast preservation is included in this chapter.

Whichever method is selected, the most suitable growth conditions for the cells must be established. It has been shown (Kirsop, 1978) that both the age of the culture and the oxygen availability during growth may substantially affect survival levels. Thus post-logarithmic cells almost always survive better than younger cells and better results with freeze-drying may be obtained with oxygen limited growth using static cultures. By contrast, survival following storage in liquid nitrogen is often higher if cells are grown aerobically using shaken cultures.

It is clear that optimum survival depends upon many factors and that selection of the method to be used will rest not only upon the specific requirements of the strain to be maintained but also upon the availability of equipment, the experience of staff and the purpose for which maintenance is required.

II.   MAINTENANCE METHODS AVAILABLE

Ten methods suitable for the maintenance of yeast cultures are described. Most have been used by the UK National Collection of Yeast Cultures (NCYC), and information relating to survival level, shelf life and the suitability of the method for different yeasts is known. Several other methods used by other culture collections are included as they may be more suitable in certain circumstances; with these methods references are given where possible, but in some cases the method has not been published.

The reader's attention is drawn to chapters 4.5 and 4.10; the methods there described as suitable for bacteria are also known to be successful for the preservation of yeasts.

The methods described are: subculturing in broth, on agar, on agar with oil, for strains with special requirements, in water; drying on silica gel and on filter paper; freeze-drying; and freezing in liquid nitrogen in ampoules or in straws.

Table I gives general information on the methods described and is intended only as a guide. As has been indicated, resistance to preservation is very strain specific and percentage survival and shelf-life figures are approximations. Convenience and cost will depend on facilities already available in laboratories.

III.   SUBCULTURING

Subculturing has been used successfully for many years and the method continues to be useful, particularly in the short term. It is simple to do, quick to carry out, and relatively inexpensive. However, it is now recognized that substantial variation may take place in strains maintained in this way over a long period. It has been found, for example, that 10% of strains showed change with regard to flocculation behaviour following maintenance by subculturing over a period of 10 years; other morphological and

Table I: Comparison of preservation methods

Method	Survival level (% population)	Shelf life (years)*	Stability†	Ease of execution†	Economy†
Subculturing					
In broth	—	0.5	1	5	5
On agar slants	<10	1	1	5	5
On agar slants with oil overlay‡	—	2	—	4	5
In water‡	—	several	—	5	5
Drying					
On filter paper (genetic strains only)	—	3–6	5	4	5
On silica gel	—	1–5	4	4	5
Freeze-drying	0.01–30	5–30	4	2	3
Freezing in liquid nitrogen					
In ampoules	20–100	>8	4	3	4/5
In straws	20–100	>8	4	3	4/5

* Known shelf life. These figures may be shown to be much higher as further information becomes available.
† Rated 1 to 5, poor to good, respectively.
‡ Methods not used by The National Collection of Yeast Cultures.

physiological properties have been found to show variation to a greater or lesser extent (Kirsop, 1974).

If strain stability is of major significance, therefore, subculturing should be minimized.

A. Subculturing in broth

1. Preparation of media   Ten-millilitre amounts of YM broth (Difco Laboratories, 0711–01) are dispensed into screw-cap McCartney bottles. The medium is sterilized by autoclaving at 121°C for 15 min. Duplicate bottles are prepared for each culture, one labelled A and the other B (see Section III.A.5).

2. Inoculation   A loopful from the B bottle of the old stock culture is transferred aseptically to each of the new bottles, the new B bottle being inoculated first.

3. Incubation   Inoculated bottles are incubated, without shaking, at 25°C for 72 h.

4. Viability estimation   Each bottle is examined visually for growth at 72 h, and macroscopic characteristics (film, ring, colour and nature of deposit) are recorded. If little growth has occurred, cultures are reincubated for a further period and examined daily. It may be necessary to provide added aeration by shaking the cultures (for *Cryptococcus*, *Rhodotorula*, *Sporobolomyces* and other aerobic genera).

5. *Storage*   The duplicate cultures are stored at 4°C. During this time the A culture is used for all operations. The B culture is retained for the sole purpose of preparing the new stock cultures and should only be opened once.

6. *Notes*   Subculturing large numbers of strains is a tedious occupation and care should be taken to relieve monotony and minimize mistakes. It is sensible, for example, to place trays of old stock cultures and uninoculated bottles on opposite sides of the operator, and to transfer old and new bottles for each strain from the trays to a different, central position before starting to subculture.

7. *Shelf life*   The majority of yeast species, other than those with unusual cultural requirements, will remain viable under these conditions for 6 months and in all probability for a much longer period. In general, fermentative species survive better than non-fermentative species, and some of the latter may need subculturing more frequently (perhaps at 2-monthly intervals).

   Species of the genera *Brettanomyces* and *Dekkera* produce large amounts of organic acids that may reduce shelf life; these strains may need more frequent subculturing or may be maintained on a medium containing $CaCO_3$ to neutralize the acids.

   It has been found that some yeast strains are sensitive to storage at 4°C. These strains must be stored at a higher temperature and, since growth will continue slowly, be subcultured more frequently.

8. *Yeasts that*   The NCYC includes yeasts of most genera and all have been
can be   maintained by this method for periods of up to 60 years. The
maintained by   reservations made above in Section III.A.7 apply, and it should
subculturing   be remembered that the method may result in considerable strain
in broth   drift. Since the reasons for loss in viability are not clearly understood, strains should be monitored in the early periods of storage in order to determine which are more sensitive.

   Genera most often found to contain strains requiring special treatment are listed in Table II, but it should be remembered that sensitive strains may occur in other genera.

B. Subculturing on agar slants

1. *Preparation*   Ten-millilitre amounts of YM agar (Difco Laboratories, 0711–
of media   02) are dispensed into screw-cap McCartney bottles, autoclaved at 121°C for 15 min, and allowed to set at an inclined angle to

Table II: Genera containing strains requiring special treatment

Genus	Treatment
Brettanomyces *Dekkera*	More frequent subculturing; $CaCO_3$ in medium
Bullera *Sporobolomyces*	More frequent subculturing; longer growth period before storage; aerobic growth
Kloeckera *Hanseniaspora*	More frequent subculturing; added vitamins may be required
Cryptococcus *Lipomyces* *Rhodotorula*	More frequent subculturing; aerobic growth
Schizosaccharomyces	More frequent subculturing

form slants. All other information given in Section III.A is applicable to this method.

2. Notes Ascosporogenous strains may sporulate on agar slant cultures when stored for prolonged periods and this may lead to strain instability. If preservation of strain characteristics is a priority, this method should not be used. If no alternative method can be used, the material for subculturing should be taken from the bottom of the slant since ascospores are generally formed at the top.

3. Shelf life Many yeasts will survive for longer periods on agar slants than in broth, particularly the non-fermentative genera (see Section III.A.7).

4. Yeasts that can be maintained by subculturing on agar slants Although this method has not been used by the NCYC for many years, all yeasts in the collection were maintained on agar slants in the past and records show a high recovery level. Non-fermentative strains often survive better on agar slants than in broth and, since many are anascosporogenous, the method may be particularly suitable for them.

C. Subculturing on agar slants with oil overlay

This method has been recommended by a number of authors (see Onions, 1971) and has the advantage of extending the shelf life of agar slant cultures.

1. Preparation of medium As for Section III.B.1.

2. Preparation of oil	B.P. medicinal oil (BDH Chemicals Ltd) is dispensed into screw-cap bottles and sterilized by autoclaving at 121°C for 15 min. Some laboratories recommend a longer sterilization time, followed by drying in an oven for several hours. Rubber washers should be removed from bottles before use.
3. Inoculation	Using a wire loop, slants are inoculated from an actively growing culture.
4. Incubation	The culture is allowed to grow at 25°C for 72 h or until good growth is obtained.
5. Oil overlay	The sterile oil is transferred aseptically to the incubated slant culture so that the oil level is 1 cm above the top of the agar slant.
6. Storage	As for Section III.A.5.
7. Notes	Care must be taken during subculturing not to let the inoculation loop splutter on flaming. Pathogens should not be maintained by this method.
8. Shelf life	The shelf life of yeasts maintained by the method described in Section III.B. may be substantially extended by covering slant cultures with a layer of sterile mineral oil and is generally in the order of 2 to 3 years.
9. Yeasts that can be maintained by subculturing on agar slants with oil overlay	The NCYC has no direct experience of this method, but it has been used by a fairly large number of laboratories over the years. There is little documentation regarding the yeast species that have been maintained successfully.

D. Subculturing strains with special properties

Strains in which specific properties must be retained, and for which freeze-drying or freezing seem inappropriate, may be subcultured in medium that will select for the required characteristic. Thus strains resistant to specific inhibitors or strains tolerant of high sugar levels may be maintained in media containing appropriate levels of inhibitor or sugar. The number of selective media of this kind is unlimited, and shelf life and suitability of the method for different yeasts will be strain specific and must be predetermined. The methods described in Sections III.A and III.B can usually be adapted to meet requirements of this kind.

E. Subculturing in water

This method has been described by Odds (1976) and used successfully by him for a number of years.

1. Method

Growth from a late-logarithmic slant culture is suspended in sterile distilled water by agitation with a wire loop. The suspended culture is transferred aseptically to a sterile container so that 90% of the volume is filled with suspension. The culture is stored at room temperature.

2. Yeasts that can be maintained in water

Candida species and a few strains from *Saccharomyces*, *Cryptococcus*, *Trigonopsis*, *Rhodotorula* and *Schizosaccharomyces* have been preserved in this way for more than 4 years by Odds (1976).

IV. DRYING

A. Paper replica method

This method has been developed by Bassel *et al.* (1977) and contributed by C. R. Contopoulou of the Yeast Genetics Stock Center, University of California, Berkeley, USA.

1. Preparation of paper replicas

Whatman No. 4. filter paper (Scientific Supplies Co. Ltd) is cut into small sections (approximately 1 cm^2) several of which (4–5) squares are placed on a piece of aluminium foil (7 × 6 cm); the foil is folded once and the packet autoclaved at 121°C for 15 min. One packet is used for each strain to be stored.

2. Inoculum

The cells to be stored are grown as a heavy patch on a plate containing a medium on which optimum growth can be attained (usually yeast extract–peptone–dextrose agar, YEPD). The plate is incubated for several days at an appropriate temperature (usually 30°C).

3. Suspending medium

Evaporated milk (any proprietary brand from food shop) is used as the suspending medium. Approximately 0.2 ml drops are transferred aseptically into sterile Petri dishes (3–4 per plate).

4. Inoculation of paper replicas

As many cells as can be retained on the blunt end of a sterile toothpick are transferred from the inoculum plate to a drop of sterile milk and mixed thoroughly. Using toothpicks or sterile forceps, the sterile filter paper sections are immersed in the cell suspension and returned to the folded aluminium foil, leaving the remaining three edges unsealed to facilitate drying. Heavy suspensions of cells improve the chances of survival over longer periods of storage.

5. Drying The packets are placed in a desiccator and allowed to dry for 2–3 weeks at 4°C. To prevent filter papers falling out from the packets and possibly causing contamination, several packets are held together securely with a paper clip.

6. Storage Once dry, the packets are removed from the desiccator and the three unsealed edges are folded. The packets are then ready for storage in a dry container kept at 4°C.

7. Revival Small pieces of the paper replicas are removed aseptically from the foil packets and are streaked across a plate containing the appropriate medium, leaving the filter paper in one corner of the plate. The plate is incubated at 30°C (23°C for temperature sensitive strains) for several days until good growth is obtained. If possible, individual clones are picked off to test for markers.

8. Shelf life Cultures are revived and replaced at intervals of 2–3 years.

9. Yeasts that may be maintained by this method The method has been used successfully for haploid and diploid strains of *Saccharomyces cerevisiae* with a wide variety of genetic markers. About 99% of several hundred strains have survived storage by this method for periods of 3–6 years.

Other species of *Saccharomyces* and other genera may not survive. Poor results have been obtained with strains of *Yarrowia lipolytica*.

B. Silica gel method

This is an adaptation of the method described in chapter 5 (Section III.A). The method has been used successfully by C. F. Roberts of the Department of Genetics at Leicester University.

1. Preparation of silica gel Purified silica gel (BDH Chemicals Ltd), mesh 6–22, is poured into McCartney bottles to a depth of about 1 cm. The gels are sterilized in an oven at 180°C for 90 min. The sterile gels are stored in a warm, dry atmosphere.

2. Suspending medium A 5% skimmed milk solution (Unipath Ltd) is prepared in distilled water. The solution is distributed in *c*. 10 ml amounts in McCartney bottles, sterilized by autoclaving at 116°C for 10 min, and stored at 4°C.

3. Inoculum   The yeast culture is grown on a YM agar (Difco Laboratories, 0711–02) slant culture for 72 h at 25°C.

4. Inoculation   Gels and milk solution are placed in a refrigerator for 24 h before
of gels   use, to become cold. The cold gels are transferred to an ice tray for inoculation. Using a Pasteur pipette, yeast cells are washed off the agar slant culture with the cold milk and a few drops of the yeast suspension are transferred to each gel. The inoculated gels are shaken to disperse the cells and are returned to the ice tray where they remain for a further 30 min.

5. Drying   The gels are kept at room temperature for about 2 weeks to dry, care being taken to screw the caps on tightly.

6. Storage   After 2 weeks, when the gel crystals separate readily, the gel bottles are transferred to an air-tight plastic container in the bottom of which is a layer of indicator silica gel (BDH Chemicals Ltd). The lids are put in place and the containers stored at 4°C. The indicator gel is checked from time to time and replenished or re-dried by heating in an oven at 180°C for 2 h, as required.

7. Revival   A few crystals are either shaken onto YM agar plates (Difco Laboratories, 0711–02) or into YM broth (Difco Laboratories, 0711–01) and incubated at 25°C for about 3 days, depending on the growth rate of the strain.

8. Shelf life   Viability using this method is strain specific and the NCYC has found some strains to be dead after 3 months and others still to be alive after 2 years' storage. Woods (1976) has recovered strains after 5 years' storage.

9. Notes   (i) It is important to keep all apparatus very cold during inoculation to minimize the effects of the heat generated when the gels are hydrated.
    (ii) The gels should not be saturated with yeast suspension; two or three drops to each gel are sufficient.

10. Yeasts   This method has been used successfully by some workers and
that may be   unsuccessfully by others. A summary of these findings is given
maintained by   below.
storage on   Of 25 species (representing 17 genera) preserved by the
silica gel   NCYC, 50% could be recovered after storage for 2 years; others were no longer viable after 1 week. Survival was strain specific and could not be related to taxonomic position. Some strains of *S. cerevisiae* that survived were no longer typical with regard to fermentation characteristics.
    Woods (1976) found that purine-requiring mutants and

polyene-resistant mutants of *S. cerevisiae* survived unchanged for 5 years; their laboratory was less successful with strains of *Candida albicans* and *Candida tropicalis*, which confirms the experience of the NCYC.

Bassel *et al.* (1977) report inconsistent results after storage of *S. cerevisiae* strains for a few years. Particularly sensitive were temperature-sensitive lethals, respiratory-deficient mutants, fatty acid-requiring strains and certain other auxotrophic mutants. They now use the method described in Section IV.A. in preference.

The NCYC has found that some strains will survive on one occasion but not on another, and it is felt that survival could be improved with further research into the effects of different parameters. The method does not offer any advantage to service culture collections, which are required to maintain a very wide spread of microorganisms, but may well be useful for more specific purposes.

V.   FREEZE-DRYING

This is a two-stage method using a centrifugal freeze-dryer, Edwards Model 2A/110 or EF03 (Edwards High Vacuum International). It can be used equally well with other centrifugal machines.

A. Centrifugal freeze-drying

1. Preparation of ampoules   Glass ampoules (FBG-Trident Ltd) are washed in detergent and rinsed in demineralized water before use. Labels for the ampoules are prepared from strips of Whatman's No. 1. filter paper (Scientific Supplies Co. Ltd). The number of the yeast and the date are printed on the label either in pencil or using a stamp with ENM quick-drying, non-toxic ink (Rexel, obtainable from stationers). The numbers are printed as near to one end of the label as possible.

The label is folded in half lengthwise (see Fig. 1) and inserted in the tube so that the writing faces outwards. The tubes are loosely plugged with non-absorbent cotton wool and those for each yeast are placed in a separate tin. The tins are either

Fig. 1   Printed label for insertion in ampoule.

sterilized in an oven or autoclaved at 121°C for 15 min. After autoclaving they are dried in an oven.

2. Inoculum

The culture to be freeze-dried is grown without aeration in YM broth (Difco Laboratories, 0711–01) at 25°C for 72 h, the amount needed depending upon the number of ampoules to be prepared. Each ampoule requires 0.1 ml of a suspension containing at least 10^6 cells ml^{-1}; in general, increasing the number of cells in the inoculum increases proportionately the number of surviving cells.

3. Suspending medium

Glucose, to a final concentration of 7.5%, is dissolved in inactivated horse serum No. 5 (Wellcome Reagents Ltd) and the mixture is sterilized by filtration and stored in McCartney bottles in a refrigerator at 4°C; sucrose or inositol may be used in place of glucose; inositol is known to provide a longer shelf life for bacteria. Other possible protective substances include skimmed milk (20%) and sodium glutamate (10%), or honey (5%) associated with skimmed milk (10%) and dextran (10%) (Berny and Hennebert, 1989) or with skimmed milk (10%) and glutamate (5%) (Berny and Hennebert, 1989).

4. Inoculation of ampoules

Equal amounts of inoculum and suspending medium are mixed aseptically in a sterile bottle, the amount prepared depending on the number of ampoules to be freeze-dried. The inoculum is not washed before use. A sterile 30-dropper Pasteur pipette is used to transfer six drops (0.2 ml) of the mixture to each ampoule, after which the cotton-wool plug is replaced. Care is taken not to touch the sides of the ampoule with the pipette, as this will distort survival calculations. The original culture is retained for a viability estimation. If this is not to be done immediately, the culture is put into the refrigerator to prevent further cell division.

5. Primary freeze-drying

The inoculated ampoules are placed immediately in the centrifuge head of the freeze-dryer with the writing on the labels facing towards the centre of the centrifuge head, so as not to be obscured by the dried yeast after centrifugation. The centrifuge head is placed on the spindle in the drying chamber. If the freeze-dryer uses phosphorus pentoxide as a desiccant the four trays are filled in a fume cupboard and placed in the drying chamber. It is important that protective clothing is worn when handling phosphorus pentoxide. The centrifuge motor and the vacuum are switched on and primary drying is continued for 3 h. At the end of this time, the centrifuge head is removed from the chamber. The yeast in the ampoules should appear completely dry.

6. Ampoule constriction

The projecting ends of the cotton-wool plugs are trimmed and the remainder of the plugs pushed halfway down the ampoules with a glass rod. The ampoules are then constricted above the level of the plug either by hand or using an ampoule constricter (Edwards High Vacuum). The phosphorus pentoxide in the freeze-dryer is replenished with fresh desiccant and returned to the chamber ready for secondary drying.

7. Secondary drying

The chamber is closed and the manifold is put in place. The constricted ampoules are placed on the manifold. Spare ampoules are used to fill any spaces. Secondary drying is generally continued overnight. Alternatively, the secondary drying may be terminated at 2 h, allowing the whole operation to be completed in one working day. After this time the ampoules are sealed under vacuum using a torch, for example that made by Dragonas Microflame. Details of the run and the number of satisfactory ampoules obtained for each yeast are recorded. One ampoule of each culture is retained for a viability estimation. The spent P_2O_5 is covered with water in a fume cupboard and discarded. If no fume cupboard is available, the trays can be left in a safe place for the desiccant to deliquesce. The trays are washed thoroughly and left in a warm place to dry.

8. Revival

Ampoules are opened as in 9(b) below, using YM broth (Difco Laboratories Ltd) to resuspend the cells. Attempts to improve survival levels by using different suspending media, times and temperatures have not been very effective, although the use of half-strength YM broth generally gave higher viable counts.

9. Viability counts

(a) *On cultures before freeze-drying.* One millilitre of the original suspension is added to 9 ml sterile glass-distilled water. Further dilutions to 10^{-7} are made. Three drops from a 30-dropper pipette (0.1 ml) of dilutions 10^{-2} to 10^{-7} are placed onto YM agar (Difco Laboratories Ltd, 0711–02), according to the Miles and Misra (1958) plate-count method. Plates are incubated at 25°C for 72 h, or longer if necessary, care being taken to keep the plates horizontal so that the drops remain discrete. Drops containing 10–30 colonies are used for estimating viability. The number of cells per millilitre inoculated into the ampoule is equal to the number of colonies in three drops multiplied by 10 times the dilution factor.

(b) *On freeze-dried cultures.* This count is carried out as soon as possible after freeze-drying. The ampoule is marked with a file at or just above the level of the cotton-wool plug. A molten glass rod is applied to crack the glass. The tip of the ampoule is removed and placed in a container for later sterilization.

YM broth (1.0 ml) is put into a sterile bijoux bottle. Using a Pasteur pipette some of the YM broth from the bijoux is added to the ampoule. The yeast is thoroughly resuspended and the whole of the suspension returned to the remaining YM broth. Using sterile forceps the label is removed from the ampoule and placed in the bijoux. A further few drops of YM broth from the bijoux may be used to wash out the ampoule again, the drops being returned to the bijoux. The suspension is a 10^{-1} dilution. Of this, 0.5 ml are transferred to 4.5 ml sterile water (10^{-2}) and further dilutions to 10^{-6} are made, plated, incubated and counted as in (a). The percentage viability of the freeze-dried culture is calculated and recorded.

10. Storage   Ampoules are stored at 4°C in the dark. Each ampoule is tested for good vacuum before storage using a high frequency spark tester (Edwards High Vacuum).

11. Notes   The NCYC has found that the degree of oxygenation provided during growth of the inoculum may affect survival. Although response to oxygen is strain specific, in general, higher survival levels are obtained following growth with restricted access to oxygen. It has also been found that growth on nutritionally poor medium before freeze-drying gave higher survival levels with sensitive strains.

12. Shelf life   Initial survival levels of yeasts are often low (see Section V.A.13), but further losses during storage are minimal. Although cultures have been preserved by the NCYC for over 30 years, and for longer periods by other laboratories, occasional strains suddenly lose viability. It is important, therefore, to monitor viability on a regular basis and to again freeze-dry strains that show a drop in viability.

13. Yeasts that can be maintained by freeze-drying   Survival of yeasts following freeze-drying is remarkably strain specific and generalizations regarding survival levels should be viewed with caution. Nevertheless, all cultures maintained at the NCYC, and covering nearly all yeast genera, have been recovered successfully, although the percentage survival of the population is generally low. Thus, the average viability figure obtained at NCYC for the genus *Saccharomyces* is 5%, for *Candida* 13%, and for *Brettanomyces* 2%. The genera in which survival levels have been particularly low are *Brettanomyces, Dekkera, Bullera, Sporobolomyces, Leucosporidium, Rhodosporidium* and occasional strains of almost all other genera.

Apart from increased levels of respiratory deficient mutants in strains known to produce petite colonies, the NCYC has detected little change in the characteristics of freeze-dried cultures, and the method has advantages with regard to strain stability.

VI.   FREEZING

A. Freezing in liquid nitrogen (ampoules)

This is a two-stage method in which cells are cooled at an uncontrolled rate to $-30°C$, allowed to dehydrate at this temperature and cooled, again at an uncontrolled rate, to $-196°C$. It has the advantage that controlled freezing rate equipment is not required.

1. Preparation of ampoules

Plastic screw-cap 2 ml ampoules are supplied sterilized from the manufacturer (Nunc, Life Science Technologies Ltd). They are labelled with the yeast number and date using a black Pentel pen (Scientific Supplies Co. Ltd), other colours having been found less satisfactory.

2. Inoculum

The culture to be frozen is grown in YM broth (Difco Laboratories Ltd, 0711–01) at 25°C for 72 h on a reciprocal shaker. Each ampoule requires 0.5 ml of a suspension containing between 10^6 and 10^7 cells ml^{-1}. Cell concentration has been found to have little effect on the percentage of cells surviving.

3. Preparation of cryoprotectant

A 10% glycerol solution is prepared, filter-sterilized, and stored in sterile screw-cap bottles [see note 9(a) below, for other suitable cryoprotectants].

4. Inoculation of ampoules

Equal amounts of inoculum and cryoprotectant are mixed aseptically in a sterile bottle. The final concentration of glycerol is thus 5%. One millilitre of the mixture is transferred to each of the sterile ampoules. Care is taken that the washer is not displaced when the cap is screwed tight as this may lead to leakage of liquid nitrogen into the ampoule; it should only be tightened until initial resistance is felt.

5. Freezing

(a) *Primary freezing.* It is very important to wear protective clothing when using liquid nitrogen refrigerators or handling frozen specimens. Ampoules are frozen to $-30°C$ by placing in a refrigerated room. If a room at $-30°C$ is not available a refrigerated cabinet or cooling bath (Camlab Ltd) may be used. If aluminium canes (Union Carbide) are to be used for the secondary freezing, ampoules may be placed on the canes at this stage. If secondary freezing is to take place in storage drawers (Union Carbide), however, the ampoules may be well spaced in wire racks for the primary freezing. The cooling rate is not critical in this method, but is probably in the region of 5°C per minute, depending upon such factors as the size of samples and containers. Cells are allowed to dehydrate at $-30°C$ for 2 h.

(b) *Secondary freezing*. The ampoules are transferred to the canisters or storage drawers of the refrigerator (Union Carbide) and immersed in the liquid nitrogen, care being taken to prevent the samples from thawing. If the distance between primary and secondary freezing containers is great, samples should be transported in a chilled Dewar flask.

6. Revival

Cultures are thawed rapidly by transferring ampoules to a water bath at 35°C and agitating them until completely thawed.

7. Viability counts

Viability counts are carried out, with appropriate adjustments, as described in Section V.A.9.

8. Storage

Ampoules are stored in the liquid nitrogen refrigerator (Union Carbide), care being taken to maintain the liquid nitrogen at such a level that ampoules are completely submerged [see note 9(c) below].

Leakage of liquid nitrogen into the ampoules may be prevented by enclosing them in polypropylene sleeves ('Nunc Cryoflex', Life Technologies Ltd), or by using the straw method described in Section IV.B. These preventative measures are felt to be preferable to storing in the vapour phase of liquid nitrogen, where the temperature is higher and may fluctuate undesirably.

9. Notes

(a) A number of other cryoprotectants have been used successfully by both the NCYC and other workers. Some have been used for a fairly wide range of yeasts, others with a few strains only. Substances used successfully include glycerol (20, 10, 5%), dimethyl sulphoxide (DMSO, 10%), glycerol plus DMSO, ethanol (10%), methanol (10%), YM broth and hydroxyethyl starch (10 and 5%).

(b) The NCYC has found that primary freezing and dehydration at -20°C, -30°C or -40°C for 1, 2 or 3 h is equally successful for the two test strains of *S. cerevisiae* used to develop the method. The intermediate protocol (-30°C for 2 h) has been adopted and has so far proved successful for all species.

(c) Cultures have been stored successfully in the vapour phase of liquid nitrogen by other workers, but as it is generally accepted that biochemical and biophysical processes may still take place, albeit slowly, at temperatures above -139°C, the NCYC prefers to store cultures at -196°C. For shorter storage periods the higher temperatures may be considered adequate.

(d) The NCYC has found that, in general, higher levels of survival are obtained from aerobically grown cultures than from those grown with limited access to oxygen. This should be compared with growth conditions found to give optimum survival in freeze-drying (see Sections V.A.2 and V.A.11).

(e) Although the NCYC has experienced only infrequent

leakage of liquid nitrogen into filled ampoules, other workers have evidence that this may occur. It can be overcome by (1) storing in the vapour phase (but see 9.c. above), (2) using the straw method (see B below) or (3) placing the ampoules in polypropylene sleeves (see 5.b. above).

10. Shelf life The survival of yeasts in liquid nitrogen for prolonged periods has not been well documented, but all evidence suggests that losses during storage are very slight. The NCYC has found no evidence to suggest that there is a drop in viability during storage for up to 8 years. In view of the high initial survival rates, shelf-life can be expected to be good.

11. Yeasts All yeast strains that have been preserved by the NCYC using
that can be this method show high viable counts. Although survival is strain-
maintained by specific, the average survival level for all genera immediately after
freezing in freezing is greater than 60%. Average survival levels are for
liquid nitrogen *Saccharomyces* 65%, for *Candida* 73% and for *Brettanomyces* 64%. Strains from the genera given in Table III have been successfully preserved by the NCYC.

Table III: Genera from which strains have been successfully preserved by the NCYC

Brettanomyces	*Metschnikowia*	*Saccharomycodes*
Bullera	*Nematospora*	*Schizosaccharomyces*
Candida	*Pachysolen*	*Sporobolomyces*
Citeromyces	*Phaffia*	*Torulaspora*
Cryptococcus	*Pichia*	*Trigonopsis*
Hansenula	*Rhodosporidium*	*Wingea*
Kloeckera	*Rhodotorula*	*Yarrowia*
Kluyveromyces	*Saccharomyces*	*Zygosaccharomyces*
Lipomyces		

B. Freezing in liquid nitrogen (straws)

This is a miniaturized adaptation of the method described in Section VI.A above. It has two advantages. First, it provides a considerable saving in storage space and, second, it provides additional security against possible contamination through leakage of liquid nitrogen into ampoules that are stored in the liquid phase. Storage in straws was first described by Gilmour *et al.* (1978), using artificial insemination straws; adaptations of the method using different kinds of straws are in use in a number of laboratories.

1. Preparation Coloured polypropylene straws (Sweetheart International Ltd)
of straws are cut into 2.5 cm lengths (Fig. 2). One end of a straw is sealed

by holding firmly in a pair of unridged forceps 1 mm inwards so that the projecting end is 1 cm from the flame of a fish-tail bunsen burner (Fig. 3). The polypropylene melts almost immediately and forms a strong seal that sets firm within a second or two.

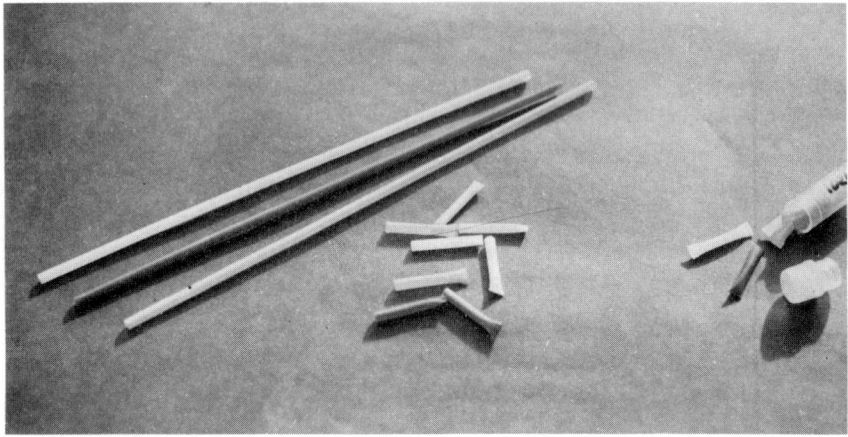

Fig. 2   Polypropylene straws: uncut (top), sealed at one end (centre), sealed at both ends (right).

The straws are placed in a glass Petri dish and autoclaved at 121°C for 15 min. The different coloured straws may be used for colour-coding yeast strains.

2. Inoculum   The inoculum is prepared as described in Section VI.A.2.

3. Preparation of cryoprotectant   The cryoprotectant is prepared as described in Section VI.A.3.

4. Inoculation of straws   Equal quantities of inoculum and cryoprotectant are mixed aseptically in a sterile bottle. A single straw is removed with forceps from the Petri dish and filled with inoculum using a Pasteur pipette. When filling it is necessary to place the end of the pipette close to the sealed end of the straw and to fill to within 3 mm of the open end. Each straw will hold about 0.1 ml of inoculum. The filled straw is then sealed at the open end as described in Section VI.B.1 above.

Six straws are placed in each plastic, screw-capped 2 ml ampoule (Nunc, Life Science Technologies Ltd) of the kind used in Section VI.A.

Fig. 3 Sealing polypropylene straws.

5. *Freezing*   Primary and secondary freezing are carried out as described in Section VI.A.5 above. Straws are frozen in the ampoules; this is easier than freezing the straws separately and transferring them, frozen, to the ampoules. No difference in survival level has been detected between straws frozen individually or in ampoules.

6. *Revival*   Using sterile forceps, a straw is removed from the ampoule and placed immediately into a screw-cap bottle containing water at 35°C. The bottle is shaken to facilitate rapid thawing. Several straws may be thawed simultaneously in this way.

7. Viable counts

Before opening the straws, the cells are resuspended by squeezing the straws several times. The straws are then wiped with 95% alcohol and one end is cut off using sterile scissors. The contents are removed using a Pasteur pipette. It may be necessary to resuspend cells further by repeated pipetting at this stage.

Two drops of the suspension (0.06 ml) are transferred to 0.54 ml sterile water to make a 10^{-1} dilution. Further dilutions, plating and counting are carried out as described in Section V.A.9.

8. Storage

Ampoules are stored as described in Section VI.A.8.

9. Notes

(a) Notes (a), (b), (c) and (d) given in Section VI.A.9 also apply to this method. It is important that the outside of the straws is kept dry during filling as wet straws do not seal well. When filling several straws with the same inoculum it is convenient to lay each straw as it is filled against a glass rod in a sterile Petri dish until all straws are filled. The straws are then sealed and put into ampoules. This is more convenient than filling and sealing each straw separately.

If the outsides of the straws are dry, they do not adhere to each other when placed in ampoules. Removal of straws from ampoules is facilitated if straws vary slightly in length. The different coloured straws can be used to colour code yeast strains to aid retrieval from the refrigerators, and it is clearly sensible to store only one yeast strain in each ampoule.

(b) Other makes of straw have been used successfully in other laboratories. The straws used for the storage of semen at artificial insemination centres have been found particularly suitable (Instruments de Médecine Vétérinaire).

10. Shelf life

The NCYC has been using this method since 1982. No loss of viability has been noticed in the test straws during this time and the high survival levels would suggest good shelf life.

11. Yeasts that can be maintained by storage in straws

Results obtained so far at the NCYC show that survival levels are equal to (and sometimes higher than) those obtained in the method given in Section VI.A. These results suggest that species that can be frozen in ampoules can be frozen equally successfully in straws.

C. Stability of genetically modified yeasts under different cryopreservation regimes

A study has been carried out (Pearson *et al.*, 1990) to determine whether freezing and thawing modified plasmid expression in *S. cerevisiae*. Freezing and thawing has been reported to induce both transient and permanent loss of expression of plasmids from *E. coli* and *Pseudomonas aeruginosa* (Calcott *et al.*, 1981; Calcott, 1982), and the induction of respiratory-deficient mutants in yeasts (Wellman and Stewart, 1973).

Current understanding is that extracellular ice formation results in an increase in the osmotic environment around the cells, causing localised increase in the media concentration around the cells and a loss of water and shrinkage (Diller and Morris, 1983; Morris *et al.*, 1988). This freeze-induced dehydration causes the wall to decrease in surface area and increase in thickness. The bilayer nature of the cell membrane dictates that normal structure can only be maintained down to the maximum packing density of the lipids. Formation of membrane invaginations allows the cells to shrink further as a direct result of water loss (Morris *et al.*, 1983a). This process appears to be reversible provided none of the membrane material is lost to the cytoplasm. On thawing, the original cell volume can be regained. The cells appear normal following a freeze–thaw cycle. Cell shrinkage during freezing is a vital component in preventing cellular damage. At rapid cooling rates there is insufficient time for the cell to shrink and, as a result, cellular electrolytes do not become concentrated and intracellular ice formation occurs. When intracellular ice formation occurs, large ice crystals form in the cytoplasm and disrupt organelles.

Three different plasmids in three yeast strains were used in the study. The plasmids used all replicated episomally and contained the *LEU2* gene of *S. cerevisiae*, a functional *E. coli* origin of replication, and differed in the fragments of the *S. cerevisiae* 2 micron plasmid (2μ) that they contained. Self-nucleating plastic straws (Cell Systems Ltd, Cambridge) were used to reduce supercooling before freezing. Different freeze profiles were applied to cultures using a controlled rate freezer (Planer, Kryo 10). A range of cooling rates were tested from 0.3°C to 64°C per minute. Normally starting at 25°C, the samples were cooled at 10°C per minute to 5°C, then at specified cooling rates to −60°C and plunge-cooled to −196°C. A modified protocol (10°C per minute to −5°C, then at specified cooling rates to −60°C and plunge-cooled to −196°C) was designed to extract the latent heat of nucleation. The cultures were removed from liquid nitrogen and warmed in a water bath at 35°C. Yeast cells were then grown up and tested for the presence of the plasmid encoded phenotype.

Both DMSO and glycerol extended the range of rates of cooling which could be used without significant loss of viability. There was a dramatic effect on cell viability with cooling rates greater than 8°C per minute. A permanent loss of yeast plasmid DNA was found to occur at the faster cooling rates. Viability was found to be more sensitive to cooling rates than to glycerol concentration. High viabilities were achieved at all concentrations of glycerol over 2.5%. In all experiments a cooling rate of 1°C per minute proved optimal for cell viability, lowest levels of petites and lowest plasmid damage. The optimum parameters were determined for the cryopreservation of plasmid-bearing *S. cerevisiae*. Maximum viability and minimum genetic damage were achieved using 15% glycerol and a cooling rate of 1°C per minute using the modified protocol.

Acknowledgement

The author acknowledges with thanks contributions from K. A. Painting and C. J. Bond of the UK National Collection of Yeast Cultures, who provided up-to-date figures on survival levels.

ALGAE AND PROTOZOA

7.1. MAINTENANCE OF ALGAE AND PROTOZOA

M. R. McLELLAN,[1] A. J. COWLING,[2]
M. TURNER[3] and J. G. DAY[2]

[1]Cell Systems Limited,
Cambridge, UK
[2]Culture Collection of Algae and Protozoa, Institute of Freshwater
Ecology, The Windermere Laboratory, Far Sawrey, Ambleside,
Cumbria, UK
[3]Culture Collection of Algae and Protozoa, Dunstaffnage Marine
Laboratory,
Oban, Argyll, UK

I. Introduction 184
II. Subculturing algae 185
   A. Strain establishment 185
   B. Media 186
   C. Light 192
   D. Temperature 192
   E. Disadvantages 193
   F. Maintenance of industrially exploited microalgae . 193
   G. Maintenance of algae for aquaculture 194
III. Subculturing protozoa 195
   A. Culture observations 195
   B. Culture vessels 196
   C. Subculture technique 196
   D. Media 197
   E. Flagellates 199
   F. Ciliates 199
   G. Amoebae 199
   H. Heliozoans 200
IV. Other preservation methods 201
   A. Cryopreservation 201
   B. Air-drying 206
   C. Freeze-drying 206

MAINTENANCE OF MICROORGANISMS 2nd Edn
ISBN 012–410351–0

I. INTRODUCTION

Strains of freshwater and marine algae and protozoa are in increasing demand for use in teaching, research and industry. Their use in classrooms and colleges for the study of basic cell biology and physiology has expanded partly because classroom use of blood and other human tissues has been discouraged, and partly due to the ease of maintenance of some algal and protozoan cultures. School or college laboratories, using window-ledges or simple artificial lighting systems, can easily maintain hardy protist strains for weeks and months. Where more sophisticated maintenance equipment is available, large numbers of strains can be supported.

Algal and protozoan culture collections of various sizes have been established around the world over the last 70 years, many offering sale or distribution of culture material. Maintenance methods vary between collections, according to economic necessity or curatorial preference. The range of maintenance methods currently available for algae and protozoa will be described here, and the virtues and drawbacks of each discussed.

For methods involving prolonged storage, it is generally agreed that the percentage recovery of a cell population following a period of maintenance should be as high as possible in order to avoid the possibility of strain loss or genetic drift. There is a general belief that in accepting methods giving low levels of viability:

(1) practical difficulties may arise in re-establishing growth if the viable cell number falls below a minimum inoculum level;
(2) pre-existing mutants may be selected.

However, there is no agreement between collections and laboratories regarding the minimum acceptable viability for the implementation of a specific maintenance method for a particular strain. Whilst some culture collections/laboratories insist on >50% recovery, maintenance methods known to give <10% or even <1% recovery are employed routinely by others (e.g. Osborne and Lee, 1975). In order that international recommendations for the acceptance of maintenance methods for protists can be made, studies are urgently required to test whether genetic drift does indeed occur in protist strains recovered from storage at low viability levels.

Another, but related, area of debate concerns the methods of viability testing used to assess population viability. Accuracy of viability testing varies according to the method chosen, and ideally a standard should be set to allow storage methods to be compared. A further discussion of viability testing is given later in this chapter.

Algae and protozoa are being increasingly exploited for industrial use and in aquaculture. Particularly where a strain is of proven value, maintenance methods must be employed to minimize risk of loss or contamination of the cell line. Recommendations for the storage of such cultures are made below.

II. SUBCULTURING ALGAE

A. Strain establishment

Following isolation from the natural environment, algal strains are maintained under largely artificial conditions of media composition, light and temperature.

The imposition of an artificial environment on a cell population previously surviving under complex, fluctuating conditions and following a seasonal life cycle inevitably causes a period of physiological adaption and/or selection, during which population growth will not occur or is very slow. For most purposes, algal cultures are maintained as uni-algal, axenic stocks, though isolation of such cultures requires considerable skill and patience (Pringsheim, 1946b; Droop, 1967). Algal strains can be maintained with bacterial, fungal, protozoan, invertebrate or other algal contaminants, where a micro-environment involving predation/symbiosis, competition and other inter-relationships will develop, affecting the physiological status of the main population. For the successful maintenance of certain species in culture, an associated bacterial flora is essential, for example *Monostroma* (Tatewaki *et al.*, 1983), where morphogenic factors produced by associated microflora are one means of ensuring normal development.

While contaminated algal cultures have previously been satisfactory for certain applications and experiments, modern experimental methods and applications demand that contaminants are not generally present, and that the taxonomy and growth characteristics of strains are defined. An important concern of many culture collections is the establishment of contaminant-free or axenic cultures. 'Cleaning' previously contaminated cultures is a skilful and time-consuming process, and could take several years in sizeable collections.

Once established, cultures enter a period of logarithmic growth, followed by a stationary phase. During the latter period, depletion of nutrients and dissolved gases and accumulation of waste products will cause deterioration and eventual loss of the culture. Therefore, it is essential that a sub-sample of viable material in late exponential or early stationary phase is transferred to fresh growth medium.

The correct maintenance of algal strains is dependant on choice of (a) media, (b) temperature and (c) light conditions. Each of these are discussed in turn below.

B. Media

The choice of culture medium for an algal strain is dependant on nutritional, morphological and taxonomic characteristics of the organism. A preliminary choice is in the selection of either liquid or solid agar-based media. Many algal strains will grow successfully both in liquid or on solid media. Development and refinement of media composition for laboratory-maintained algal cultures has been the object of research for several decades, resulting in many different media 'recipes' being reported in the literature and being used in different laboratories (Turner and Droop, 1978).

Media can be classified as being defined or undefined (Turner and Droop, 1978). Defined media, which are often essential for nutritional studies, have constituents that are all known and can be assigned a chemical formula. Undefined media, on the other hand, contain one or more natural or complex ingredients, for example agar, liver extract and seawater, the composition of which is unknown and may vary. Defined and undefined media may be further subdivided into freshwater or marine media.

In choosing or formulating a medium, it may be important to decide whether or not it is likely to promote heavy bacterial growth. Richly organic media should be avoided unless the algae being cultivated are axenic. For contaminated cultures, mineral media should be used. These may contain small amounts of organic constituents, such as vitamins or humic acids and, in either case, provide insufficient carbon for contaminating organisms to outgrow the algae. Some examples of media that accord with this scheme of classification are given below.

1. Jaworski's medium: defined freshwater mineral medium (Thompson et al., 1988)

Stock solutions:

		per 200 ml
(1)	$Ca(NO_3)_2.4H_2O$	4.0 g
(2)	KH_2PO_4	2.48 g
(3)	$MgSO_4.7H_2O$	10.0 g
(4)	$NaHCO_3$	3.18 g
(5)	FeNa-EDTA	0.45 g
	Na_2-EDTA	0.45 g
(6)	H_3BO_3	0.496 g
	$MnCl_2.4H_2O$	0.278 g
	$(NH_4)_6Mo_7O_{24}.4H_2O$	0.2 g

		per 200 ml
(7) Vitamin B$_{12}$		0.008 g
Vitamin B$_1$		0.008 g
Biotin		0.008 g
(8) NaNO$_3$		16.0 g
(9) Na$_2$HPO$_4$.12H$_2$O		7.2 g

Final solution:

Stock solutions	(1)–(9) 1 ml of each
Glass-distilled water	1.0 litre

2. S66.
Defined
freshwater
organic
medium

S66 medium (modified)

	per litre
NaCl$_2$	0.3 g
MgCl$_2$.6H$_2$O	0.05 g
KCl	0.0075 g
CaSO$_4$.2H$_2$O	0.01 g
Glycylglycine	0.5 g
Glycine	0.25 g
KNO$_3$	0.1 g
K$_2$HPO$_4$	0.01 g
Vitamin B$_{12}$	100.0 ng
Vitmin B$_1$	50.0 µg
Citric acid	0.04 g
Trace mineral solution	10.0 ml

Make up to 1 litre with glass-distilled water. Adjust pH to 8.0.
Trace mineral solution (Droop, 1961)

	per litre	final conc. per litre
FeSO$_4$.7H$_2$O	250 mg	500.0 µg
MnSO$_4$.H$_2$O	15.5 mg	50.0 µg
ZnSO$_4$.7H$_2$O	2.2 mg	5.0 µg
CuSO$_4$.5H$_2$O	1.9 mg	5.0 µg
CoSO$_4$.7H$_2$O	0.24 mg	500.0 ng
MoO$_4$Na.2H$_2$O	0.12 mg	500.0 ng

3. SES (*Soil*
extract
medium with
salts)
undefined
freshwater
mineral
medium
(*Thompson* et
al., *1988*)

Stock solutions:

	per litre
(1) K$_2$HPO$_4$	1.0 g
(2) MgSO$_4$.7H$_2$O	1.0 g
(3) KNO$_3$	10.0 g

Final solution:

Soil extract stock	100.0 ml
Stock solutions (1)–(3)	20.0 ml

Make up to 1 litre with glass-distilled water.
Soil extract stock:

Mix one part air-dried, sieved soil with two parts distilled water. Adjust to pH 8 with NaOH or HCl and autoclave for 1 h at 15 lb/in^2 pressure. Decant or filter the supernatant.

4. 'Euglena		per litre
Gracilis	Sodium acetate (trihydrate)	1.0 g
medium'	'Lab Lemco' powder (Unipath L29).	1.0 g
(EGM):	Tryptone	2.0 g
undefined	Yeast extract	2.0 g
freshwater	$CaCl_2$ stock solution	10.0 ml
organic medium	Non-nutrient agar (optional)	15.0 g
(Thompson et		
al., *1988)*	$CaCl_2$ stock solution	1.0 g

5. ASP2:		per litre
defined	NaCl	1.8 g
marine	$MgSO_4.7H_2O$	0.5 g
mineral	KCl	0.06 g
medium	Ca (as chloride)	10 mg
(Provasoli et	$NaNO_3$	5 mg
al., *1957)*	K_2HPO_4	0.5 mg
	$Na_3SiO_2.9H_2O$	15 mg
	TRIS	0.1 mg
	Thiamine HCl	0.05 mg
	Nicotinic acid	0.01 mg
	Ca pantothenate	0.01 mg
	p-aminobenzoic acid	1.0 µg
	Biotin	0.1 µg
	Inositol	0.5 mg
	Folic acid	0.2 µg
	Thymine	0.3 µg
	Na_2-EDTA	3.0 mg
	Fe (as chloride)	0.08 mg
	Zn (as chloride)	15.0 µg
	Mn (as chloride)	0.12 mg
	Co (as chloride)	0.3 µg
	Cu (as chloride)	0.12 µg
	B (as H_3BO_3)	0.6 mg
	H_2O	100 ml
	pH	7.6–7.8

6. S88 and		per litre
vitamins:	NaCl	16.0 g
defined	KCl	0.4 g
marine	$MgSO_4.7H_2O$	2.5 g
organic	$CaSO_4.2H_2O$	0.5 g
medium	$SrCl_2.6H_2O$	6.5 mg
(Turner,	$AlCl_3.6H_2O$	0.25 mg
1979)	RbCl	0.10 mg
	$LiCl.H_2O$	0.05 mg

		per litre
	Kl	0.025 mg
	KBr	32.5 mg
	Glycine	500.0 mg
	KNO_3	100.0 mg
	K_2HPO_4	10.0 mg
	Na_2-EDTA	50.0 mg
	$FeSO_4.7H_2O$	2.5 mg
	$MnSO_4.4H_2O$	203.0 µg
	$ZnSO_4.7H_2O$	22.0 µg
	$CuSO_4.5H_2O$	19.6 µg
	$CoSO_4.7H_2O$	2.38 µg
	$NaMoO_4.2H_2O$	1.26 µg
	Vitamin B_{12}	100.0 ng
	Vitamin B_1	50.0 µg

Adjust pH to 8.0 using 1.0 M NaOH/1.0 M HCl.
Vitamins are made as additions after the dry mixture has been dissolved, prior to adjusting pH and autoclaving.

7. Guillard's medium (Thompson et al., 1988)

Trace element solution

	per litre
Na_2-EDTA	4.36 g
$FeCl_3.6H_2O$	3.15 g
$CuSO_4.5H_2O$	0.01 g
$ZnSO_4.7H_2O$	0.022 g
$CoCl_2.6H_2O$	0.01 g
$MnCl_2.4H_2O$	0.18 g
$Na_2MoO_4.2H_2O$	0.006 g

Vitamin solution

	per litre
Vitamin B_{12}	0.5 mg
Vitamin B_1	100 mg
Biotin	0.5 mg

Final solution

	per litre
$NaNO_3$	0.075 g
$NaH_2PO_4.2H_2O$	0.00565 g
Trace element solution	1 ml
Vitamin solution	1 ml

Adjust pH to 8.0 using 1.0 M NaOH/1.0 M HCl. Make up to 1 litre with filtered natural seawater.

8. SNA		per litre
(Seawater	'Ultramarine' Synthetica sea salts	17.5 g
nutrient	(or filtered natural seawater)	(500.0 ml)
agar):	Nutrient agar (Unipath)	28.0 g
undefined	Glass-distilled water	to 1.0 litre

marine
organic   Variation
medium   Follow the recipe for SNA but add 35 g 'Ultramarine'
(Thompson et   Synthetica sea salts (Waterlife Research Industries Ltd) (or
al., *1988)*   1000 ml filtered natural seawater) (i.e. double quantity) and
                 add 35 g NaCl.

9. Hetero-		per litre
trophic growth	Artificial sea salts (Werner Aquatechnic)	16.0 g
medium for	Nutrient broth	8.0 g
Tetraselmis	Glucose	10.0 g

suecica (Day
et al., *1991)*

10. Comments   (a) *Rationalization and standardization.* Culture collections have
attempted to rationalize the number of media recipes, and to
standardize recipes for algal strain maintenance. In particular,
the use of undefined biphasic media (soil/water mixture) is
declining, due to lack of reproducibility in media batches and
occasional contamination of the media from soil samples.

(b) *Photoautotrophic algae.* These require a medium containing
a nitrogen source (generally nitrate), phosphate, major inorganic
trace elements, and sometimes organic micronutrients. A typical
medium for photoautotrophs is 'Jaworski's medium' (see 1,
above). Obligate heterotrophs often require an external carbon
source, generally supplied as acetate or glucose. Some crypto-
phytes, volvocalean flagellates and euglenoid flagellates require
such a carbon source. For example, *Euglena gracilis* and
Chlamydomonas sp. can be grown effectively on EGM (see 4,
above). Nitrogen-fixing cultures of cyanobacteria (usually hetero-
cystous forms) should generally be maintained on media free of
or deficient in combined nitrogen. Otherwise, strains may lose
heterocyst function or structure (Castenholz, 1988).

(c) *Diatoms.* An external silica source is required by diatoms,
generally supplied as Na_2SiO_3 as in ASP2 (see 5 above).

(d) *Heterotrophic growth.* For certain industrial applications, it
is necessary to grow algae under heterotrophic conditions. Maintenance of industrial starter cultures is by heterotrophic or
mixotrophic means; a typical heterotrophic growth medium for
Tetraselmis suecica is given in Section 9 above.

(e) *Water.* If natural waters are not used to prepare freshwater
and marine media, then the constituents can be dissolved in

distilled water. Filtered natural seawater is an excellent basis for a maintenance medium; a disadvantage is that such water can demonstrate seasonal variation in its ability to support growth. Localized chemical pollution in water samples may also occur and go undetected. A complete list of algal medium recipes used by a modern culture collection is included in Thompson *et al.* (1988).

(f) *Preparation of media.* Media may be prepared by combining concentrated stock solutions which are not combined before use, to avoid precipitation or contamination. Another practicable way of preparing artificial defined marine media is to mix the ingredients together and to dry them prior to long-term storage. Constituents are added sequentially to distilled water, smallest quantities first, to form a solution and finally a stiff slurry. Prior to the addition of the major salts, the pH is adjusted to between 4 and 5. The mixture is then transferred to a clean desiccator of suitable size and the final additions are made. The well-stirred slurry is vacuum-dried, a process which usually takes 4 or 5 days for 2 or 3 kg material. It should be stirred frequently, especially at the beginning and end of the process. The dry mixture is then stored with calcium chloride as a desiccant. Enough medium for 100 litres can be conveniently prepared in this way and, even complete with vitamins, can be stored for some years if kept dry.

(g) *Culture vessels.* These are usually borosilicate glass conical flasks or test tubes for liquid culture, and test tubes for agar culture. Vessels are capped by non-absorbent cotton-wool plugs, which will allow gas transfer but prevent entry of microbial contaminants. A more efficient, though costly, way of capping vessels is to use Silicosen rubber bungs [Jencons (Scientific) Ltd], which also allow efficient gas transfer; these are re-usable and easy to handle. Sterilization is by autoclaving at 121°C (15 lb/in^2), heat-labile compounds being added to sterilized media by filter sterilization (Nucleopore Filters, 0.2 µm diameter).

(h) *Solid media.* Usually these are prepared using 1.0–1.5% agar, tubes being rested at an angle of 30° during agar gelation to form a slope that increases the surface area available for growth.

(i) *Subculturing.* Generally, subculturing is performed using aseptic microbiological techniques; bunsen flame, microbiological loops (steel or plastic disposable) and glass or plastic sterile pipettes are required.

Intervals between routine subcultures vary between 2 weeks and 6 months, depending on conditions of light and temperature, and whether agar or liquid medium is used. Sterility testing of axenic cultures should be made at each transfer. A successful protocol for routine maintenance of axenic algae is to keep a set of three or four cultures of each strain. If the subculture interval is 2 months, a new culture is established from the culture which is 2 months old. A sterility test can then be carried out on the culture which is 1 month old: a loopful of material is inoculated into a richly organic liquid medium and this is incubated in the dark at 20°C to allow heterotrophic contaminants to grow up. If the sterility test medium is clear and uncontaminated after incubation, the culture tested can be used as an inoculum. If contamination has occurred, an older uncontaminated culture is used as an inoculum.

C. Light

Artificial light from fluorescent tubes (warm white or cool white) is the preferred means of energy supply to photoautotrophic and mixotrophic algal cultures. Illumination is typically at 100 μmol m^{-2} s^{-1}, although taxonomic groups vary in optimal light intensity required for growth. Many strains of blue-green algae require lower light intensity (25 μmol m^{-2} s^{-1}), whilst diatoms thrive under light conditions of 200 μmol m^{-2} s^{-1}. In large collections, an appropriate way of providing different levels of illumination is by shading glass shelves using mesh or paper. Day–night light cycles can be controlled using a 24-hour timer, 16 h on and 8 h off being the most common cycle used. A typical growth room is illustrated in Figure 1.

Fig. 1   A typical culture room, containing algal cultures on slopes and in liquid medium.

D. Temperature

As with light intensity, the optimal temperature for growth varies between algal strains. In large culture collections, therefore, it is difficult to provide optimal temperatures for all strains. It is often desirable to maintain cultures at a sub-optimal growth temperature, to prolong intervals between subculturing. Generally cultures are maintained at 15°C, with minimum and maximum temperatures being 10 and 20°C, respectively. The use of lower temperatures to maintain cultures has been suggested; Umebayashi (1972) reported that several strains of marine diatoms can be maintained for many months at 5°C without subculture, providing cells were given short light periods several times per day.

E. Disadvantages

While serial subculture as a maintenance method for algal strains has the advantage that strains are immediately available for distribution, the method contains a number of serious disadvantages. Growing cultures are at risk of contamination from airborne or mite-carried bacterial, fungal or yeast cells, particularly during subculturing routines. Assuming vessel seals are efficient, contamination should not occur in standing cultures. Infection of axenic cultures should be avoided, as axenic starter cultures are difficult to obtain and may be of economic value. A second disadvantage is that subculturing is time-consuming, expensive and laborious. Expenses include media, glassware, sterile equipment, growth room, power and man hours.

A particular problem is encountered with the subculturing of diatoms, due to their decrease in size during successive divisions (Jaworski *et al.*, 1988). This eventually leads to an inability of some cultures to divide, eventually leading to the loss of the culture (over a period of years) unless auxospores can be produced.

A final disadvantage of subculturing is that a continual selection pressure is applied to strains maintained in an artificial growth environment, which may well result in genetic changes over long periods.

F. Maintenance of industrially exploited microalgae

Microalgal biotechnology has to date involved a relatively small number of organisms (Table I) and techniques of culture maintenance currently used are largely traditional, with primary cultures maintained by serial transfer on agar slopes incubated in temperature controlled incubators under a 16 h light/8 h dark cycle. In general, the final production stage of photoautotrophic processes are non-axenic; however, care is taken to ensure culture purity is maintained to the 2–20 litre stage in inoculum build-up. A number of processes have been developed on the basis of growing microalgae under heterotrophic conditions. Starter cultures are maintained under heterotrophic or more commonly mixotrophic conditions; this ensures that any enzymes associated with substrate uptake/utilization are constitutative within the primary inoculum, also culture densities are

Table I: Microalgae currently produced on an industrial scale

Organisms	Product
Spirulina platensis	Biomass, health food and pigments
Chlorella spp.	Biomass, health food
Haematococus pluvialis	Astaxanthin
Dunaliella bardawil	β-carotene
Tetraselmis suecica	Aquaculture feed
Cyclotella cryptica	Aquaculture feed

much greater than autotrophic culture, giving a larger initial inoculum size.

It must be stressed that, to date, the majority of microalgal products are produced using wild-type strains. At present, a great deal of strain development is being carried out, resulting in valuable patented strains. It is unlikely that serial transfer will prove to be satisfactory for the maintenance of such production strains or mutants, whether generated by conventional mutagenesis or genetic engineering. Alternative means of maintaining cultures are discussed in Section IV.

G. Maintenance of algae for aquaculture

Algal cultures are used extensively as hatchery feed for juvenile fish and crustaceans, and for mature shellfish. The principles of algal culture used by the aquaculture industry are little different from those employed by research workers and culture collections. Scaling up presents some problems, however, especially regarding contamination. Ideally, small starter cultures should be unialgal and axenic, and should be maintained apart from large-scale feed cultures.

Cultures larger than 5 litres are difficult to maintain under axenic conditions because of difficulties with autoclaving or filter-sterilizing large volumes. In practice, optimal growth conditions of light, temperature and gas supply for the algal strains are employed, in order that the relative size of contaminant populations is minimized.

Natural waters generally form the basis of bulk media for aquaculture, and pre-filtering is essential (Helm *et al.*, 1979). Chemical sterilization using hypochlorite solution, followed by neutralization with sodium thiosulphate, and passage through ultra-violet sterilization systems may also be used in appropriate situations (Baynes *et al.*, 1979).

Hundred-litre plastic bags and tanks of various materials are used as vessels for growing algae on a mass scale indoors (De Pauw and Pruder, 1986; Fig. 2). In such hatcheries, artificial

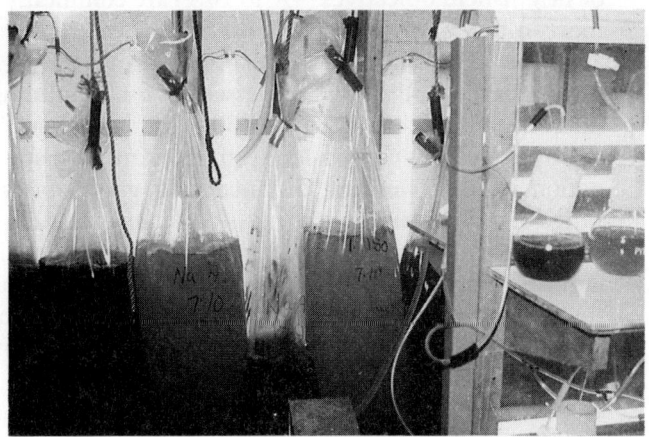

Fig. 2   Algal cultures grown in 100 litre bags. (Photo courtesy of Golden Sea Produce Ltd)

indoor lighting systems comprising of banks of fluorescent tubes are employed. Some hatcheries, particularly in climates where sunlight quality and duration are assured, grow algae used as feed outdoors under non-axenic conditions.

Choice of media varies from fully defined media, through natural water with trace elements and vitamins added, to natural water with no additions where the nutrient quality can be assured.

Continual on-site production of strains is subject to contamination and is costly in labour and equipment. Some hatcheries are replacing live cultures with heterotrophically-grown, spray-dried algal cells, for example *Tetraselmis suecica* and *Cyclotella cryptica* (Algal 161™, Algal 262™, Cell Systems Ltd). While viability is not retained in such cultures, following rehydration cellular integrity is maintained, allowing ingestion by particle feeders.

III. SUBCULTURING PROTOZOA

Heterotrophic protozoa can be cultured using a variety of methods, with varying degrees of success. In this section, techniques employed for the routine maintenance of protozoan isolates in batch culture by serial subculture are discussed. Other methods of preservation are detailed in Section IV.

A. Culture observations

An inverted microscope provides a versatile means of examining a range of protozoa in laboratory cultures. Low-power dark-ground or bright-field objectives (4× to 10×) are commonly used and are most often suited to the observation of larger ciliates and amoebae. The higher magnifications provided by 20× to 40× phase-contrast objectives are essential for examining cultures of microflagellates and small (<30 μm) protozoa. The distinct advantage of a typical modern inverted microscope is its combination of long working distance with good optics and a wide range of objectives. This permits direct examination with a single instrument of protozoan cultures maintained in a wide variety of culture vessels. Nevertheless, a hand lens is often quite adequate for assessing culture densities of tube culture populations of ciliates, and a binocular stereomicroscope can be put to good use for general observations of strains in culture.

B. Culture vessels

'Pyrex' rimless culture tubes (150 × 16 mm, *c*. 15 ml capacity) are suitable for the culture of many free-swimming protozoa (e.g. most bacterivorous ciliates, euglenid flagellates) and are ideal for áxenically-grown strains where media can be dispensed and autoclaved in individual tubes. Cotton-wool plugs have now been superceded by steel, aluminium or polypropylene caps. Disposable sterile plastic tubes (usually screw-capped) of varying capacities (10 — 20 ml) can be an improvement (i.e. longer culture 'shelf-life') for strains maintained in axenic media at sub-optimal growth temperatures (<20°C). Sterile plastic Petri dishes are a suitable alternative for many culture strains, especially raptorial ciliates, suctorians, heliozoans and some others that require frequent addition of food protozoa.

Sterile plastic tissue culture flasks (e.g. Bibby, Nunc) are well suited to the maintenance of many ciliate, flagellate and amoebae species, particularly marine isolates that can be maintained by the simplest of methods: in either artificial or natural seawater with added wheat or rice grains (sterilized by boiling in water). Sterile mediaware is convenient for routine cultivation purposes, but can be more expensive than equivalent re-usable glass vessels.

C. Subculture technique

It is good practice to adopt aseptic techniques (use of laminar flow cabinets, sterilized media, pipettes, etc.) wherever possible, regardless of the culture condition of the maintained strain. This reduces the risk of both airborne contamination and any cross-contamination from other microorganisms maintained in the same laboratory.

There is no general rule governing the frequency and type of inoculum transfers needed when subculturing protozoa. The subculture inoculum itself can often be the most critical consideration for the successful and continued maintenance of healthy culture lines. Only experience will show whether subcultures of any one particular strain are better initiated from a few cells, from a comparatively young culture or from a larger inoculum transferred from a well-established older culture. The former method is more generally applicable for 'purer' types of culture (axenic, monoxenic, dixenic) than for the cruder agnotobiotic or polyxenic cultures, where simultaneous transfer of sufficient food organisms (bacteria, other protozoa, etc.) to initiate new cultures is essential.

Liquid inocula may be transferred by sterile loop or straight wire (e.g. small flagellates), micropipette (small protozoa generally) or standard Pasteur (glass) or plastic disposable pipette (larger protozoa). New tube cultures are initiated simply by pouring (aseptic technique) from the chosen inoculum tube into tubes containing fresh media. Techniques for strains maintained by surface growth of trophic cells on agar plates are described below for Gymnamoebae.

D. Media

The so-called 'biphasic' soil/water (S/W; see Table II for media used for protozoa, and abbreviations) tube-culture method, devised by Pringsheim (1946a) is still one of the most convenient methods for maintaining a large range of protozoan isolates — most notably euglenid and chrysomonad flagellates; hymenostome ciliates, scuticociliates and some spirotrichs (e.g. *Spirostomum*). Culture tubes are prepared simply by adding approximately 3–4 cm air-dried soil (preferably an untreated garden loam at neutral pH) and a barley grain to each tube, and filling (to

Table II: Media used for maintenance of protozoa

Abbreviation	Media
AS	*Amoeba* saline
ASW	Artificial seawater
MW	Mineral water
MP	Modified Pringsheim's solution
NSW	Natural seawater
PM	Polytoma medium
PC	Prescott's and Carrier's
PJ	Prescott's and James'
PPY	Proteose peptone yeast extract
S/W	Soil/water ('biphasic')

Full details in Thompson *et al.* (1988).

three-quarters full) with water, then autoclaving. Leedale (1967) recommends the use of soils with high clay content for biphasic euglenid cultures, whilst Pringsheim's original study suggested adding a diversity of supplementary material (e.g. starch, cheese, etc.) to satisfy more closely the needs of specific flagellate strains.

Axenic media suitable for routine strain cultivation have been developed for amoebae (*Acanthamoeba, Naegleria*), ciliates (*Paramecium, Tetrahymena, Uronema, Parauronema*) and flagellates (*Astasia, Chilomonas, Euglena, Polytoma, Polytomella* and *Peranema*). Complex media formulations suitable for the axenic cultivation of groups such as choanoflagellates (Gold *et al.*, 1970) and others can be found in the literature, but they have rarely been used for the long-term maintenance of strains in service collections.

Soil extract media, sometimes supplemented with wheat grains, are commonly used for chysomonad and bodonid flagellates, although better and more consistent growth of these and other groups can often be achieved using plant material infusions. The most versatile and consistently balanced media can be made from commercially available dried cereal leaf preparations such as 'Cerophyl' (International Marketing Corporation) or an equivalent (e.g. Sigma dehydrated cereal leaf preparation C7171).

Inorganic salt solutions such as PJ, AS and MP have been developed for those protozoa that grow less well in the media described above but persist as healthy populations in such solutions when regularly (usually weekly) provided with washed food organisms. Bottled mineral water (MW) is often a more suitable alternative to these, in having a suitably 'pure' and consistent mineral composition. There are various brands commercially available; 'Volvic' natural mineral water' (Perrier UK Ltd) has been used successfully for this purpose, and sterilization by autoclaving does not impair its usefulness.

Refrigerated storage of media both prevents evaporation and prolongs shelf life. Care should be taken to allow temperature acclimation of media to live cultures or vice versa; likewise, media stocks should be checked for occasional contamination. Contamination of plates, for example, may not be immediately obvious in stocks maintained at refrigerator temperatures and removed for immediate use. Wherever possible, all culture media for protozoa should be sterilized by autoclaving or filtration.

References giving details of isolation and identification and culture methods for freshwater protozoa are given by Finlay *et al.* (1988). A useful and comprehensive specialist account, which includes cultivation details for specific protist groups, is given by Margulis *et al.* (1989). Formulations for selected, commonly used media are given below and in Thompson *et al.* (1988).

General maintenance conditions. The laboratory conditions to which protozoan batch cultures are subjected, and their consequent growth response, will be determined largely by temperature, medium type and food availability. Convenient incubation temperatures lie between 15 and 20°C for the majority of laboratory cultivated strains; where possible, reduced temperatures (e.g. 7°C for cyst formers) can significantly reduce the frequency of subculture required.

It should be emphasized that the best solution will inevitably be a compromise between the necessity to have healthy, dense cultures available at any one time, and to reduce as far as possible the work involved in culture maintenance. The aim therefore is to reduce the frequency of subculture, and this often involves selecting sub-optimal conditions for protozoan growth.

E. Flagellates

Soil extract or cereal infusion media in tissue culture flasks or Petri dishes, with or without added wheat grains, are suitable media components for the culture of flagellate strains. Many euglenids (e.g. *Astasia, Chilomonas, Distigma, Gyropaigne, Hyalophacus, Khawkinea, Menoidium, Parmidium, Rhabdomonas, Rhabdospira*) may be maintained successfully using biphasic tube cultures, but they will also grow well in Petri dishes or tissue-culture flasks in inorganic salt solution with added wheat grains. Further details of methods for flagellate heterotrophs are given in Cowling (1991).

F. Ciliates

Plant infusions (including Cerophyl-based media) support good growth of many bacterivorous ciliates. Soil-extract media are also suitable. Both culture types may require the addition of either one or more cereal (wheat or rice) grains and/or other protozoa as food to produce dense cell populations. Many formulations for axenic ciliate media have been developed; one of the more commonly used is PPY for the culture of *Tetrahymena*.

G. Amoebae

Many strains of Gymnamoebae may be grown on non-nutrient agar (NNA) or malt-yeast extract agar (e.g. MY75S) plates with a suitable food bacterium (cultured separately on nutrient agar) added by spreading bacteria across the agar surface (Page, 1988). New subcultures are initiated by preparing fresh plates with bacteria onto which is placed a small (<5 mm diameter) block of agar excised from a healthy inoculum source plate. *Escherichia coli* is commonly used; plates are maintained at 20°C and subcultured at intervals varying from 10 days to 6 weeks. Page (1988) gives further details of such culture methods and media for many amoebae genera.

Some Gymnamoebae genera (e.g. *Naegleria, Acanthamoeba, Paratetramitus*) can be conveniently stored as cysts on agar slopes at a reduced temperature (*c.* 7°C), where they will remain viable for 6 months or longer. Other protozoa that readily encyst (particularly terrestrial ciliates and flagellates) may similarly be maintained as old or dried cultures, excystment and production of active (trophic) forms being induced by the addition of liquid and/or food bacteria, etc. Induction of amoebae excystment is carried out by placing excised agar blocks (with a dense surface layer of cysts) onto fresh plates streaked with food bacteria such as *E. coli*. Alternatively, cysts or trophic amoebae may be washed from agar plate surfaces using a stream of suitable medium [e.g. AS or diluted (75%) seawater, depending on the strain] and transferred as a cell suspension by pipette. It is necessary to point out that care must be taken when dealing with cultures of *Naegleria* and *Acanthamoeba* as these genera contain species which are pathogenic to man. Strict observation of general laboratory safety procedures (use of containment facilities, masks, autoclaving of waste cultures, etc.) is required to eliminate the potential risk of infection through eyes, nose or mouth. Production of aerosols of any kind when subculturing should be prevented.

The larger amoebae (e.g. *Amoeba, Chaos*) can be maintained in tissue-culture flasks or dishes of similar capacity (30–40 ml) containing PC or MP to which appropriate food organisms are added as washed dense cell suspensions. Suitable food organisms for these genera are:

Amoeba:	*Tetrahymena*
Chaos:	*Colpidium*
Polychaos:	*Chilomonas*

Culture populations of testate amoebae such as the euglyphids *Euglypha, Trinema, Assulina* and *Corythion*, and others such as *Arcella*, may be maintained on CP (Cerophyl-Prescott) agar plates spread with food bacteria and overlain with a shallow liquid medium layer such as Cerophyl or soil extract (Cowling, 1986). Larger forms such as *Difflugia* and *Netzelia* may require the addition of food algae such as *Chlorogonium*, as well as the provision of fine sand particles as shell-building material.

H. Heliozoans

Autoclaved MW is a convenient, simple medium in which good growth of heliozoans (e.g. *Acanthocystis, Actinophrys,*

Actinosphaerium, Raphidiophrys) can be obtained using frequent (weekly or bi-weekly) provision of food ciliates (i.e. *Tetrahymena* and *Colpidium*).

IV. OTHER PRESERVATION METHODS

A. Cryopreservation

Cryopreservation is the maintenance of viable cells and tissues at low, sub-zero temperatures. In order for the method to be successful the stabilization temperature must not exceed $-130°C$, the threshold temperature for ice re-crystallization (Mazur, 1966).

Several configurations of cryostorage units are available: mains-driven freezers capable of achieving and maintaining $-135°C$ (e.g. Gallenkamp Supercold 135), are suitable for the storage of cryopreserved cultures. More commonly, liquid nitrogen is used as a convenient refrigerant, with full immersion of frozen samples ($-196°C$) or suspension of samples in the gaseous phase above ($-135°C$). Storage units are available in a range of sizes with inventory systems for efficient organization of cryopreserved cultures. Many have auto-filling options and alarm systems to detect unfavourable temperature excursions (e.g. Planer Products Ltd, Statebourne Cryogenics Ltd). An overwhelming advantage of cryopreservation as a maintenance method is that, once organisms are cryopreserved by a proven method giving high cell recovery, viability is independent of storage time. Thus, some algal strains have been recovered with no significant decline in viability 13 years after initial storage (McLellan, 1989). A possible disadvantage to the method, especially where the number of strains to be maintained is relatively low, is that in addition to relatively expensive storage equipment, specialized freezing equipment and consumables are required.

1. Equipment for Freezing   Algal and protozoan cultures are frozen either in sterile 2 ml plastic screw-top ampoules (Nunc, Life Science Technologies Ltd) or in 0.25–5.00 ml straws (IMV Ltd). The former can be labelled by a suitable permanent pen marker or colour-coded using a colour insert located in the tube cap. The latter are available in a range of colours for coding and can be labelled either using a permanent marker or printed using a Domino Jet Printer System (Domino Printers Ltd). Sealing of straws can be achieved using heat sealing, or sealing compound (IMV Ltd). Technology for the scale-up of cryopreservation of cell lines has recently been introduced. Bags (capacity 25 ml) lined with a nucleating agent, giving acceptable cooling profiles, are available (Xybags™, Cell Systems Ltd) and may be useful for the bulk cryopreservation of seed-stock for algal fermentations. Some algal strains require precise regulation of cooling rate during freezing in order for viability to be maximized following thawing. Therefore, controlled-rate freezers providing linear cooling rates are needed (e.g. Planer Products Ltd, Cryo-Med Ltd). For the majority of algal strains, however, the optimal cooling rate need only be met between +20°C and −30°C, followed by plunging to −196°C. In these instances a −30°C immersion bath (e.g. Fryka Kaltetechnic) and a liquid nitrogen dewar are sufficient for effective cryopreservation. For thawing samples, a 40°C water bath is all that is required.

2. Pre-culture: growth temperature

The freezing resistance of some freshwater (Morris, 1976), and marine (Ben-Amotz and Gilboa, 1980) algae, and ciliated protozoa (Polyansky, 1963) increases with a decrease in growth temperature of the culture. By transferring an early stationary phase culture to a hardening temperature (4–5°C) for several days prior to freezing, recovery following cryopreservation can be significantly improved. For example, 4°C-hardened *Chlorella emersonii* demonstrated 50% recovery after 20 days culture post-thaw, compared with 0% in non-hardened cells frozen by the same method (Leeson *et al.*, 1984).

3. Pre-culture: age of culture

Cultures in exponential phase are in general more sensitive to freezing stress than stationary phase cultures (Morris, 1978). Early to mid stationary cultures should be frozen rather than cultures in late stationary phase, since viability can decline and physiological stress can result from nutrient depletion.

4. Pre-culture: nutrient limitation

This has been shown to improve freezing tolerance in some algal strains, e.g. limitation of nitrate (Morris *et al.*, 1980), bicarbonate (Ben-Amotz and Gilboa, 1980) or silica (McLellan, 1989).

5. Cryo-protective additives

Some algal strains, e.g. *Chlorella vulgaris*, can be cryopreserved in the absence of cryoprotective additives, providing the cooling rate is optimized. However, for effective cryopreservation of most protistan strains, it is essential that a cryoprotectant be added. These are generally dimethyl sulphoxide (DMSO), glycerol or methanol. The former additives can be sterilized in culture media by autoclaving; however, methanol must be filter-sterilized, because it is denatured by heat exposure.

6. Cooling rate

It is now well documented that algal strains have an optimal cooling rate for maximum cell recovery on thawing (Morris,

1978). Faster rates cause reduced viability due to an increased probability of intracellular ice. Slower rates result in a similar decline in viability due to freezing-induced hypertonic stress (Mazur, 1966). Optimal cooling rates have been determined for a wide range of algal species, with many strains, particularly Chlorococcales, having an optimum cooling rate of 10–15°C per minute in the presence of cryoprotectant. With such strains a simple freezing method is available using a −30°C freezing bath and liquid nitrogen dewar. The experimental scheme is outlined in Figure 3.

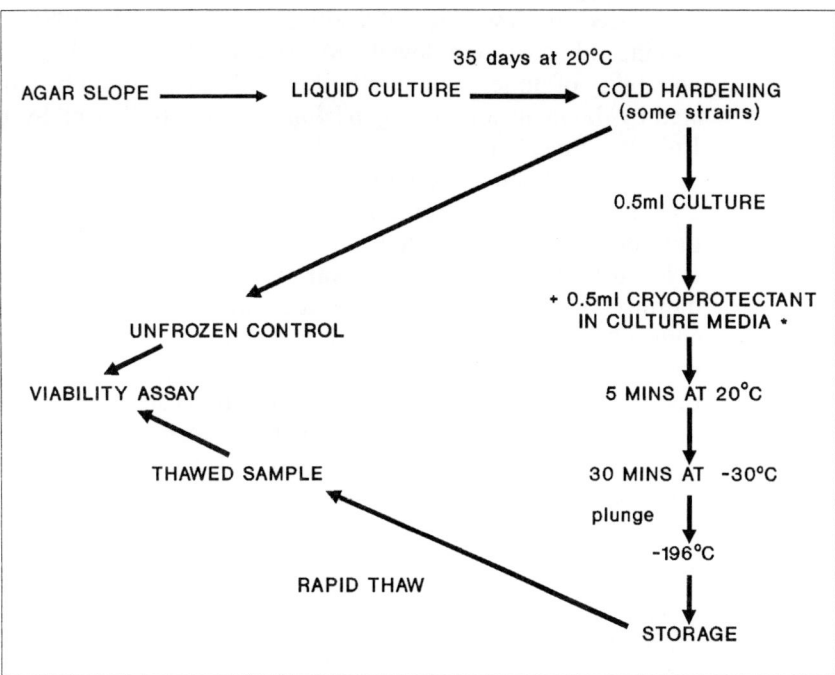

Fig. 3   A two-step cryopreservation method used for some algal strains.

7. *Notes*   It should be stressed that the technique illustrated in Figure 3 is not applicable to all protist strains. Many algal strains require carefully controlled cooling regimes, for example the desmid *Cylindrocystis brebissonii* (Morris *et al.*, 1986). To achieve these cooling rates, a controlled-rate freezer is required.

For many algal and most protozoan strains, successful cryopreservation strategies with high rates of population recovery remain to be developed. Recently developed methods for the cryopreservation of non-parasitic protozoa, particularly *Naegleria*, result in significant rates of recovery following freezing. For other protozoan species, zero or low recovery values have been obtained, and novel cryopreservation strategies must be devised.

Innovation and improvement of cryopreservation methods can best be achieved by understanding and applying the basic principles of cryobiology.

The literature dealing with the attempted cryopreservation of protozoa has indicated that many protist strains suffer 'cold-shock', a form of low temperature injury not linked to ice formation but caused by a rapid reduction in temperature. *Amoeba* (McLellan *et al.*, 1982), *Blepharisma* (Giese, 1973), and *Chlamydomonas* (Morris *et al.*, 1983b) are genera in which cold-shock injury has been demonstrated. To avoid cold-shock injury the cooling rate should be reduced, since cold-shock stress is minimized at slow rates of cooling (Leeson *et al.*, 1984), and specific additives employed (Morris, 1987). In addition, the possibility of undercooling of bathing solutions should be eliminated by the inclusion of Xygon™ or Cryo-seeds™ (Cell Systems Ltd) or by mechanical seeding. A particularly useful method for the establishment of cryopreservation procedures is to use a cryomicroscope (McGrath, 1987; Fig. 4). This allows visualization of cells at controlled rates of cooling with or without cryoprotective additives. Survival of cells can be assessed using motility or vital stains, and successful methods scaled-up in controlled rate freezers for viability testing. A cryomicroscope study of *Hartmanella*, a small freshwater Gymnamoeba, suggests that cooling rate and cryoprotectant addition can be optimized to avoid cold-shock injury, intracellular ice formation and freeze-induced hypertonic damage (Brown and McLellan, in prep.).

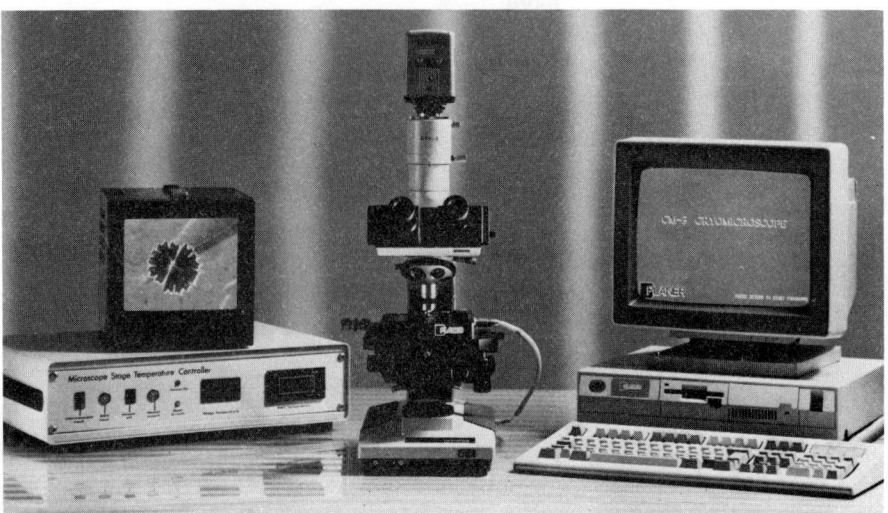

Fig. 4   A typical cryomicroscope system, Planer CMS. (Photo courtesy of Planer Products Ltd)

These results should enable a method for the recovery of *Hartmanella* at high recovery values to be developed in batch.

8. Assays of viability

One of the problems in comparing methods of storage and assessing the relative efficiency of each has been the variation in methods of assessing cellular viability. Fundamental assessment of viability (i.e. whether or not regrowth of cultures occurs) must be considered as crude measures of population survival. Quantitative assays based on dye exclusion by the cell membrane, or motility, can provide useful estimates of survival, but generally over-estimate recovery potential. For example, with *Tetrahymena pyriformis*, many cells that are motile immediately on thawing do not retain the capacity to divide (Osborne and Lee, 1975). Therefore, assays based on cell division should be used whenever possible. Two such assays are described below.

(1) *Colony formation in agar.* Many Protista will form discrete colonies on agar. However, with motile cells it is essential to assay viability within the agar. This latter technique is described in detail below.

Logarithmic dilutions of the cells are prepared either in their growth medium or a balanced salts solution. One millilitre of the appropriate suspension is pipetted into a sterile Petri dish, after which an appropriate nutrient solution containing agar (*c.* 1.5% w/v) at 40–42°C is poured in. The cells are dispersed through the agar by gentle agitation and, when set, the plate is inverted and incubated under suitable cultural conditions. For cell types requiring long-term culture before the development of visible colonies, dehydration of the agar may be reduced by sealing the Petri dish with 'Clingfilm'. For the assay of non-motile Protista on the surface of the agar, the technique is similar to that described above, except that the diluted cell suspension is applied to the surface of a prepared, solidified agar plate from which surface condensation has been removed. The cell suspension is distributed over the agar by using either the conventional spread plate method or the Miles and Misra (1958) technique.

In the agar system, a colony can arise from either a single cell or from a cell clump. Because many unicellular organisms in liquid culture tend to form aggregates, it is essential to estimate cellular multiplicity in order to determine accurately the total cell number. The frequency of cellular aggregation is determined microscopically from three samples, at least 500 cells being counted for each treatment. The multiplicity of the sample is then defined as:

$$\text{multiplicity} = \frac{\text{No. of cells}}{\text{No. of groups}} .$$

Thus the Total no. of cells = No. of colonies × multiplicity, and % viability

$$= \frac{\text{Total number of cells before treatment}}{\text{Total number of cells after treatment}} \times 100.$$

This technique can be 'miniaturized', using microwell plates in which four sets of five serial dilutions can be made on a single plate.

(2) *Most Probable Number Technique.* With organisms that will not readily form colonies in or on agar, viable cell numbers may be determined by most probable number techniques in liquid media. For details of these methods as originally developed using tubes of media, see Swaroop (1938). To economize, methods using micro-titre plates have been developed for the assay of ciliates (Heaf and Lee, 1971) and may be successfully applied to other protists.

9. *Thawing rates*   In all studies in which the effects of rate of warming from −196°C on the survival of Protista have been examined, rapid rates gave maximum survival. For practical purposes, this can be achieved by immersing tubes or straws in a water bath at 40°C. This results in a warming rate of 90°C per minute in tubes and 250°C per minute in straws.

B. Air-drying

Many microalgae live in extreme environments where they are periodically subjected to severe drought. As an evolutionary response to such stress, many strains produce resting stages that can remain dormant until their environment improves. This ability can be exploited for the maintenance of microalgae having resistant resting stages. Air-dried soil samples containing cysts of *Haematococcus pluvialis* have been shown to be capable of generating a new culture after 27 years of storage (Leeson *et al.*, 1984); furthermore, cultures of *Nostoc muscorum* and *Nodularia harveyanna* have been regenerated from dried soil samples that were stored for up to 79 years (Roach, 1920). Vegetative cells may also be preserved by air-drying: the ability of dried, alginate-immobilized *Chlorella emersonii* to recover photosynthetic capacity and the ability to reproduce after 1 year storage has been reported (Day *et al.*, 1987). However, there is little information regarding the percentage recovery that can be expected using this method.

C. Freeze-drying

Freeze-drying has in general been found to be an unsatisfactory method for the preservation of eukaryotic microalgae, with very low survival rates on lyophilization and further reduction in viability on prolonged storage (Holm-Hansen, 1964, 1967; Day, 1987). Cyanobacteria, in contrast, have been demonstrated to be amenable to conventional freeze-drying techniques with good recovery after up to 5 years' storage (Holm-Hansen, 1973). This method of preservation has been adopted by a number of culture collections for the preservation of selected cyanobacterial strains.

At present, the methodology for freeze-drying algae is crude, and further research is warranted in order that freeze-drying methods for algae and protozoa can be developed. The profound advantage in the development of a successful freeze-dry protocol would be that strain maintenance could be at room temperature, with a minimum requirement for specialized maintenance effort or equipment.

Acknowledgement

The authors wish to acknowledge the contribution from Mitzi De Ville of CCAP, Windermere, UK.

ALGAE AND PROTOZOA

7.2. MAINTENANCE OF PARASITIC PROTOZOA BY CRYOPRESERVATION

E. R. JAMES

Medical University of South Carolina,
Charleston, South Carolina, USA

I. Introduction 209
   A. Effect of cooling 210
   B. Cryoprotectants............................. 211
   C. Storage and re-warming...................... 212
II. Preparation 218
   A. Capillary tubes 218
   B. Ampoules 219
   C. Latent heat of fusion....................... 220
   D. Cryoprotectants............................ 220
III. Cryopreservation methods 221
   A. Slow cooling............................... 221
   B. Rapid cooling 222
   C. Stepped cooling............................ 223
   D. Storage and documentation 223
   E. Thawing 224
IV. Survival assays................................ 225

I. INTRODUCTION

It was recognized long ago (Coggeshall, 1939) that the techniques in existence for laboratory maintenance of parasitic protozoa were frequently tedious, expensive and precarious. Irreversible biological changes often occurred in the material being maintained, and strains could easily be lost by a decline in virulence through successive passage or the unanticipated death of an infected host or contamination of a culture. Cryopreservation overcomes these problems and is now an indispensible tool in any work with parasitic protozoa.

The methods used to cryopreserve parasitic protozoa are many and various, reflecting the different requirements of the different

MAINTENANCE OF MICROORGANISMS 2nd Edn
ISBN 012–410351–0

species of protozoa, as well as the funds available to the various workers and the local conditions in which they have to operate. Many of the techniques described in the literature may, because of these specific limitations, be sub-optimal. It cannot be stressed strongly enough that a set of conditions that leads to good survival of one species will very possibly be wholly inappropriate for another species. Different life-cycle stages or strains of the same parasite may similarly require different methods. A particular cryopreservation protocol should only be treated as a basic guideline.

Survival of parasitic protozoa following cryopreservation has usually been extremely low, even in many of the more recent studies in which considerable attention has been given to the addition and removal of cryoprotectants and the cooling and warming procedures have been carefully defined. Often this has not mattered too much, since the parasites have been capable of reproducing rapidly in culture or in their host following injection. However, where survival levels are particularly low, this will impose a selective pressure on the parasite population. It is imperative, therefore, to use a technique that aims to give the highest possible survival rate of the particular parasite species.

A. Effect of cooling

During slow cooling, water in the medium in which cells are suspended is converted into ice and the salts become concentrated into the remaining unfrozen fraction. This creates an osmotic imbalance across the cell's outer membrane and so the cell dehydrates. The sensitivity of different cell types to the concentrated levels of extracellular solutes produced during freezing, to the different cryoprotectant compounds and to dehydration, varies considerably.

If the rate of cooling is faster than the rate at which the cell can dehydrate to maintain its water equilibrium, the cell's intracellular contents will freeze. The size of the intracellular ice crystals which form depends on the cooling rate, the crystals being smaller when faster rates are used. At very fast rates, the ice crystals may be so small that cell damage is avoided.

Very rapid cooling rates coupled with low storage temperatures and very fast warming rates can inhibit ice crystal formation and encourage the water to vitrify. It is almost impossible to achieve the very rapid cooling needed to vitrify pure water, however; the presence of dissolved solutes and the addition of cryoprotectants considerably slows the cooling rate required for vitrification of culture media and cell contents. Except for *Naegleria* (Luyet and Gehenio, 1954) though, the use of very rapid cooling has not been well explored for parasitic protozoa.

Between the dehydration/osmotic-induced slow-cooling injury and the intracellular ice-induced injury of fast cooling there usually lies a zone of survival for a particular type of cell at an intermediate cooling rate. This intermediate cooling rate typically becomes slower as increasing concentrations of cryoprotectant are used. Some cells that are particularly small in size and/ or have a surface membrane that is highly permeable to water will dehydrate sufficiently, even at cooling rates of several hundred or even several thousand degrees per minute. Mammalian red cells survive well at very fast cooling rates and, hence, the best cooling rates for many of the intra-erythrocytic parasites tend to be in this range. However, many other parasitic protozoa appear to survive best with slower cooling rates in the region of 1°C per minute. Table I gives a synopsis of those published techniques that have attempted to define an optimum set of conditions for the cryopreservation of parasitic protozoa.

Two-step cooling, in which cells undergo slow or rapid cooling to a particular intermediate temperature, and after a holding period, further rapid cooling to the storage temperature, can reduce some of the potentially damaging effects of ice and solutes. The cells become partially dehydrated during the slow cooling step, or at the intermediate holding temperature if the initial cooling step is rapid. Then, during the second, rapid cooling step to the storage temperature, either the intracellular contents are vitrified or the ice crystals, if formed, may be too small to be damaging. The intermediate temperature is critical: at temperatures above the optimum the cells are not dehydrated enough, so that large internal ice crystals form during the second cooling step. At holding temperatures below the optimum, or with extensive periods at the intermediate temperature, the cells become too dehydrated and/or the concentration of solutes is damaging. For parasitic protozoa, two-step cooling was first applied to trypanosomes (Cunningham *et al.*, 1964). It has also been used successfully for *Plasmodium* (Wilson *et al.*, 1977) and for *Entamoeba* (Farri *et al.*, 1983).

B. Cryoprotectants

Cryoprotectants, by definition, enhance the survival of cryopreserved cells. They perform this function by one or more of several different mechanisms. Cryoprotectants depress the freezing point of water so that at any temperature a smaller fraction of the suspending medium will be composed of ice. The dissolved solutes will thus be more dilute and exert less of an osmotic or chemical stress on the suspended cells. This reduction in salt/solute toxicity is thought to account for a large proportion of the protective action of cryoprotectants. Many cryoprotectants contain OH groups, which interact with water by hydrogen bonding and may help to stabilize liquid water and inhibit ice-crystal formation. Additionally, cryoprotectants may increase the viscosity of the suspending medium, often quite considerably. The likelihood of a sample vitrifying depends on the cooling rate used and the viscosity of the suspending medium and intracellular contents. The addition of a highly viscous cryoprotectant such as glycerol may encourage vitrification of samples cooled at even relatively slow rates. Many cryoprotectants will also protect against ionizing radiation, probably through their ability to scavenge free radicals. There is growing evidence that free radicals may be produced during freezing and may contribute to cell damage.

C. Storage and re-warming

During final storage, cell deterioration and death is generally more likely to occur at high temperatures, particularly above $-136°C$, the devitrification temperature, although under certain conditions, particularly using glycerol, extended storage at higher temperatures is feasible. Yaeger (1988) recently reported successful recovery of *L. donovani* cryopreserved in 10% (v/v) glycerol at $-75°C$ for 23 years. At lower temperatures, almost indefinite storage is possible. Slow warming is usually damaging to cells but can sometimes give higher levels of survival than rapid warming if the cooling rate used was slow. However, optimum survival following rapid cooling is obtained by rapid warming; this indicates that at least some of the damage associated with intracellular ice formation occurs during warming. If small ice crystals do form during rapid cooling, these will grow into larger damaging crystals during slow warming.

Special care should be taken, when transferring samples between storage vessels, to protect them from warming by even a few degrees. Also, when removing individual samples for thawing, other samples in the same rack or canister should be prevented from warming.

Few of the many techniques described for the cryopreservation of parasitic protozoa have been derived by careful empirical research. The trypanosomes have been most extensively studied; following investigations by Cunningham Lumsden and others, techniques of slow cooling (between 0.7 and 5°C per minute from 0 to $-60°C$) and fast warming (approximately between 8 and 2000°C per minute from $-60°C$) were developed which gave survival levels that were little different from unfrozen controls when assayed by an infective dose test (ID_{50} ml^{-1}) (Webber *et al.*, 1961). The slow cooling technique described in Section III is based on this method.

Table I: Synopsis of published successful techniques for cryopreservation of parasitic protozoa giving good levels of survival.

Parasite (stage)	Container, volume	Additive:* type, concentration	Equilibration Time	Equilibration Temp.	Cooling rate	Thawing rate	Dilution procedure	Reference
Aegyptianella pullorum	5 ml plastic tube	Glycerol, 10%	30 min	4°C	fast to −196°	Fast in 37°C water bath	–	Raether and Seidenath (1972, 1977)
Anaplasma marginale	6 ml serum vials, 4 ml	DMSO, 2 M(?)	–	–	fast to −196°C	Fast in water bath at 37°C	–	Love (1972)
Babesia bigemina	Plastic straw, 0.5 ml	DMSO, 14.2%	30 min	0°C	82°C min⁻¹ to −60°C, fast to −196°C	Fast, in water bath at 40°C (930°C min⁻¹)	–	Dalgliesh and Mellors (1974)
Babesia bovis (culture and extracellular forms)	Plastic cryotubes, 1.0 ml	PVP, 10%	–	–	20°C min⁻¹ to −70°C, slow to −196°C	Fast, in water bath at 37°C	1:50 in Pucks Saline G or VYM medium	Palmer *et al.* (1982), Vega *et al.* (1985)
Babesia rodhaini	Plastic straws, 250 µl	DMSO, 10.7%	–	0°C	100°C min⁻¹ to −100°C, then fast to −196°C	Fast, in water bath at 40°C (930°C min⁻¹)	–	Dalgliesh *et al.* (1980)
Cryptosporidium (oocysts)	1.5 ml plastic ampoule, 1 ml	DMSO, 5–10%	–	–	0.3°C min⁻¹ to −80°C, fast to −196°C	Moderate, in water bath at 0°C	1:50 in PBS	Rossi *et al.* (1990)

Table I. *Continued.*

Parasite (stage)	Container, volume	Additive:* type, concentration	Equilibration Time	Equilibration Temp.	Cooling rate	Thawing rate	Dilution procedure	Reference
Dientamoeba fragilis	Screw-cap vials, 0.5 ml	DMSO, 2.75%	30 min	25°C	5.5°C min^{-1} to −6°C, 3.5°C min^{-1} to −60°C fast to −196°C	Fast, in water bath at 37°C	–	Dwyer and Honigberg (1971)
Eimeria (sporozoites)	Glass ampoule, 1 ml	Glycerol, 7.5%	20 min	20°C	1°C min^{-1} to −70°C, fast to −196°C	Fast, in water bath at 37°C	None	Norton *et al.* (1968)
Eimeria (sporocysts)	Glass ampoule, 1 ml	DMSO, 7.5%	15 min	20°C	1°C min^{-1} to −70°C, fast to −196°C	Fast, in water bath at 37°C	–	Norton and Joyner (1968)
Entamoeba histolytica (trophozoites)	1 ml glass ampoule, 0.5 ml	DMSO, 7.5%	15 min	37°C	1°C min^{-1} to −100°C, fast to −196°C	Fast, in water bath at 37°C	–	Farri *et al.* (1983)
Giardia intestinalis	Plastic ampoule, 0.5 ml	DMSO, 7.5–10.0% in TYI-S-33 medium	–	0°C	1°C min^{-1} to −25°C then 5°C min^{-1} to −196°C	Fast, in water bath at 37°C	1:20 in TYI-S-33 medium	Phillips *et al.* (1984)
Giardia lamblia	–	DMSO, 6.5% in TPS-1 medium	–	–	1.19°C min^{-1} to −40°C, then rapid to −196°C	Fast, in water bath at 37°C (400–500°C min^{-1})	–	Lyman and Marchin (1984)

Table I. *Continued.*

Parasite (stage)	Container, volume	Additive:* type, concentration	Equilibration		Cooling rate	Thawing rate	Dilution procedure	Reference
			Time	Temp.				
Leishmania tropica var. *major*	Corked glass tube, 0.5 ml	DMSO, 10.65%	–	0°C	1.9°C min^{-1} to −65°C, fast to −196°C	600–800°C min^{-1} in water bath at 37°C	>1:10 in 10 mM glucose saline	Callow and Farrant (1973)
Leishmania spp.	Glass capillary, 25 μl	Glycerol, 7.5%	10 min	20°C	0.7°C min^{-1} to −60°C, rapid to −196°C	Fast, in air at 20°C	–	Lumsden *et al.* (1973)
Naegleria	Glass ampoule, 0.5 ml	DMSO, 5.0%	30 min	20°C	1.3°C min^{-1} to −55°C, fast to −196°C	Fast, in water bath at 35°C	–	Simione and Daggett (1976)
Plasmodium chabaudi (trophozoites)	Glass capillary, 25 μl	Glycerol, 10%	20 min	20°C	Plunge into LN at 3600°C	Rapid, in water bath at 37°C	15% (w/v) glucose	Mutetwa and James (1984a,b)
P. falciparum (trophozoites)	Glass capillary, 25 μl	Glycerol, 10%	20 min	20°C	Capillary in glass tube at 300°C min^{-1} to −196°C	Rapid, in water bath at 37°C	15% (w/v) glucose	Mutetwa (1983), Mutetwa and James (1985)
P. galinaceum (trophozoites)	Glass capillary, 25 μl	Glycerol, 10%	20 min	20°C	1°C min^{-1} to −80°C, rapid to −196°C	Rapid, in water bath at 37°C	15% (w/v) glucose	Mutetwa (1983), Mutetwa and James (1985)

Table I. *Continued.*

Parasite (stage)	Container, volume	Additive:* type, concentration	Equilibration Time	Equilibration Temp.	Cooling rate	Thawing rate	Dilution procedure	Reference
P. berghei (sporozoites)	Glass tube, 1 ml	HES, 10% serum, 50%	—	0°C	50°C min^{-1} to −80°C, then to −196°C	Fast, water bath at 37°C (300°C min^{-1})	—	Leef *et al.* (1979), Hollingdale *et al.* (1985)
Theileria parva	Glass capillary 100 μl, or tube 2.5 ml	Glycerol, 7.5%	—	—	Slow (1°C min^{-1}?) to −80°C, then to −196°C	Fast, in water bath at 37°C	1:10 or 1:50 in serum + 7.5% glycerol	Cunningham *et al.* (1973)
Toxoplasma gondii	Glass ampoule, 2 ml	Glycerol, 5%	—	—	0.3°C min^{-1} to −79°C	Fast, in water bath at 37°C	None	Eyles *et al.* (1956)
Toxoplasma gondii	Glass ampoule, 1 ml	Glycerol, 5%	—	—	1°C min^{-1} to −160°C	Fast, in water bath at 37°C	1:5 in MEM	Bollinger *et al.* (1974)
Trichomonas vaginalis	Glass tube, 1 ml	DMSO, 7.5%	0 min	25°C	In −30°C freezer 90 min, then to −75°C freezer	Fast, in water bath at 37°C	—	Miyata (1975)
Trichomonas vaginalis	Glass ampoule, 2 ml	DMSO, 5%	20 min	20°C	1°C min^{-1} to −35°C, fast to −196°C	Fast, in water bath at 45°C	1:20 in Diamond's medium	Ivey (1975)

Table I. *Continued.*

Parasite (stage)	Container, volume	Additive:* type, concentration	Equilibration Time	Temp.	Cooling rate	Thawing rate	Dilution procedure	Reference
Trypanosoma brucei spp.	Glass capillary, 25 µl	Glycerol, 7.5%	10 min	20°C	0.7°C min⁻¹ to −60°C, rapid to −196°C or −140°C	Fast, in air at 20°C	−	Lumsden *et al.* (1973)
T. b. congolense	Screw-cap vial, 250 µl	Glycerol, 12%	30 min	24°C	1−2°C min⁻¹ to −59°C, rapid bath to −196°C	Fast, in water bath at 41°C	−	Diffley *et al.* (1976)
T. cruzi	1 ml ampoules, 0.5 ml	DMSO, 10%	5 min	−	3°C min⁻¹ to −60°C then to −196°C	Fast, in water bath at 37°C	1:5 in saline	Ribiero dos Santos *et al.* (1978)
T. cruzi	Plastic ampoule, 2 ml	Glycerol, 7%	6 min	20°C	2°C min⁻¹ to −60°C, fast to −196°C	Water bath at 35°C	1:9	Engel *et al.* (1980)

* Abbreviations: DMSO, dimethyl sulphoxide; HES, hydroxyethyl starch; TPS-1, tryptone-panmede-serum (see Diamond *et al.*, 1978); TY1-S-33, trypticase-yeast extract-iron-serum (see Diamond *et al.*, 1978); VYM, Vega y. Martinez (see Vega *et al.*, 1985); LN, liquid nitrogen; MEM, minimal essential medium; PBS, phosphate-buffered saline; PVP, polyvinylpyrrolidone; all cryoprotectant concentrations are % v/v.

Note: The corresponding cooling rate values are expressed in units of $^{\circ}C\,min^{-1}$.

II. PREPARATION

Although different species of parasites require different cryo-preservation techniques for optimal survival, the overriding consideration for choosing or developing a particular technique is probably the apparatus available for storage. Because the use of ampoules restricts the upper range of cooling rates that can be used to about 200–400°C per minute depending on the material from which the ampoule is constructed and the volume being preserved, many workers cryopreserve all their organisms in glass capillary tubes. However, there are certain advantages and disadvantages both to using capillary tubes and ampoules.

Glass capillaries and ampoules are flame-sealed and thus they can be used for the human pathogenic protozoa and for material that must be sterile. The disadvantage of glass is that it can shatter, so there may be a risk of infection. Also, if improperly sealed, liquid nitrogen may enter the tube or ampoule during storage and this will almost certainly cause it to explode upon thawing. Some cryoprotectants, such as sucrose and certain polymers, have lower coefficients of expansion than glass and thus can cause glass ampoules, and particularly glass capillaries, to shatter during cooling. Polypropylene ampoules (e.g. Nunc cryotubes, Life Science Technologies) are favoured by many workers because they do not crack or fracture. Indeed they have become the method of choice for storage in many applications. However, although the silicone rubber rings are seated within the caps and the manufacturers say leakage cannot occur, even with the caps screwed down tightly they can very occasionally leak, and thus sterility may be more difficult to maintain. Neverthe-less, these tubes are safer because, if liquid nitrogen does penetrate during storage, the consequences are less severe.

In recent years there have been improvements to the plastic straws used for storage of spermatozoa, and these may be suitable for certain protozoa. Again, entry of liquid nitrogen into one of these straws is dangerous as the plastic plugs or ball bearings can be blown out of the ends at high velocity during thawing. Polypropylene drinking straws have been successfully used for other organisms (see ch. 6) and, as plugs are not used, the hazard is minimal.

A. Capillary tubes

The glass capillary tubes should be cleaned in chromic acid (three parts saturated potassium dichromate solution to one part sulphuric acid). Detergents do not generally produce adequate cleaning. Recently introduced cleaning agents such as RBS 35 (for example Pierce) may be substituted however. The tubes are rinsed three times in tap water followed by three times in demineralized or double glass-distilled water prior to use. After drying in a warm oven capillaries then should be stored in test tubes (e.g. 125 × 16 mm) sealed with an aluminium foil cap. If sterilization of the tubes is required, autoclaving at 121°C is suggested (Lumsden et al., 1973) since capillaries sterilized by dry heat do not fill so easily with the parasite suspensions. Final drying of autoclaved tubes can be carried out in a warm oven.

If a large number of samples in capillaries is to be cryopreserved in one batch, it is convenient to have pivotted tube holders constructed which can accommodate 25 or more tubes each. Tubes of 100 mm length and 1 mm external diameter are commonly used (light wall type, Plowden and Thompson Ltd), but the size is not especially critical. The suspension of parasites and cryoprotectant is drawn into a Pasteur pipette, which is used to fill about 20 mm of the length of each tube as rapidly as possible. The individual capillaries (or the tube holder) are then tilted so that the suspensions run to about the centre of the length of the tubes. Each tube is then sealed with a microburner [e.g. Soudagaz S, Baird and Tatlock (London) Ltd], starting with the end through which the suspension was introduced, this being more difficult to seal. The seal can then be checked if the tubes are tilted into a vertical position. Improper sealing can be identified by the suspension running down inside the tubes, and these tubes should be resealed or discarded. The other ends of the tubes are then sealed with the microburner, avoiding the formation of a terminal bubble or bead if possible. This can be achieved by pushing the tube 3–4 mm into the burner flame to heat the air in the end of the tube, and then by sealing the extreme tip. Perfect sealing of the tubes is important to conserve sterility and when storing the tubes in liquid nitrogen.

With experience, the approximate time taken to add cryoprotectant to the suspension of parasites and to fill 25 tubes is in the region of 10 minutes; this is thus the minimum equilibration time in the cryoprotectant before commencing the cooling procedure.

B. Ampoules

Ampoules are often more convenient containers than capillaries for some applications and some parasites. Larger volumes can be frozen and stored more easily and, being larger than capillaries, they are also easier to label, which is an aid to inventory control. The main disadvantage is that the contents of ampoules cannot be cooled very quickly; however, the optimum cooling and warming rates of most parasitic protozoa appear to lie within the range attainable with these containers, so this consideration may not apply. For parasites for which the optimum cooling rate lies above that attainable with ampoules, some workers nevertheless prefer to use ampoules and overcome the reduced survival by using much larger volumes of material to initiate their cultures or to infect their laboratory animal maintenance hosts. This practice should be discouraged, primarily because it may place a selective pressure on the parasite population.

It is convenient first to draw out the neck of the glass ampoule in a flame and then score and break off the end so that the parasite suspension has to be introduced using a syringe and needle. This helps to minimize the risk of contamination of the material and reduces the amount of heat required for flame sealing. For ampoules made of polypropylene with screw caps and a silicone rubber sealing ring (e.g. Nunc cryotubes, Life Science Technologies Ltd (Nunc)) the caps should be screwed on tightly prior to final storage to minimize the chance of liquid nitrogen entry.

C. Latent heat of fusion

During cooling, the temperature fluctuations that accompany the evolution of latent heat of fusion are potentially damaging to living cells. This can be overcome with samples in screw-cap ampoules by carefully 'seeding' the sample with a small ice crystal once it has been cooled to just below the freezing point. The crystal is deposited using the tip of a Pasteur pipette containing a sample of the suspending medium frozen in liquid nitrogen. This operation is, however, impossible with flame-sealed glass ampoules or capillaries. Seeding in these containers can sometimes be achieved by sharp mechanical agitation or flicking or, alternatively, by holding a small piece of metal pre-cooled in liquid nitrogen against the wall of the container. Some commercially available, programmable freezing machines incorporate methods to overcome the evolution of latent heat of fusion and produce near perfect linear cooling rates right through this critical region.

D. Cryoprotectants

Cryoprotectants can be added in neat undiluted form to the suspension of parasites, but neat glycerol at 0°C is very viscous and so accurate measurement of small volumes is difficult. If added neat, then a table of the glycerol volumes relating to specific suspension volumes to produce the desired final cryoprotectant concentrations should be drawn up. An accurately calibrated 0.25 or 1.00 ml syringe, or an adjustable automatic pipette, is used to add the required volume of glycerol or other additive to the measured volume of parasite suspension, which is then mixed using a Pasteur pipette to draw and expel the suspension three or four times. The same pipette can then be used to fill the capillary tubes or ampoules.

Dimethyl sulphoxide is a solid at 0°C and reacts exothermically when added to water, so it should first be made up as a double-strength solution in the suspending medium and added in an equal volume to the parasite suspension. Most workers also prefer to use this method of addition with the other cryoprotective compounds.

The preferred temperature of addition of the cryoprotectant is approximately 20°C (room temperature), but frequently 0°C or 37°C are used. Strict control of the duration of exposure to the additive prior to freezing should also be maintained.

III. CRYOPRESERVATION METHODS

The method used to achieve a specific cooling rate is less important than knowing what cooling rate is produced. Programmable freezing machines provide excellent control of the cooling and warming rates, and are particularly useful when experiments are being conducted with new species or strains of parasite to identify the optimum requirements. Many laboratories use these machines routinely, but they can be relatively expensive and consume fairly large amounts of liquid nitrogen. However, improvised inexpensive methods have been devised by many workers to achieve slow, relatively well-controlled rates.

A. Slow cooling

One simple method of achieving a cooling rate of just under 1°C per minute over the range 0 to −60°C for capillaries is to pack them into a plastic holder (see D, below) in an aluminium tube which in turn is placed in an insulated jacket. This is then placed in a refrigerator set at −70°C or below and left overnight. This method has been used for many years by the WHO Collaborating Centre for Trypanosomiasis and Cryopreservation of Protozoa at the London School of Hygiene and Tropical Medicine, and has been described in detail in the book *Techniques with Trypanosomes* by Lumsden *et al.* (1973). The heat capacity of the aluminium tube containing the capillaries is approximately 210 J and the insulating jacket is made of expanded polystyrene with 25 mm thick walls.

When access to a low-temperature refrigerator is not possible, the capillary holder or ampoule can be surrounded with plasticine to a depth of approximately 10 mm and this plug placed into the neck of a liquid nitrogen dewar vessel. Experimentation is required with this method to determine the precise positioning of the plug and the thickness of its walls, in order to give the desired cooling rate. A more sophisticated and relatively cheap variable-rate apparatus operating on the same principle is commercially available (Union Carbide, Biofreezer, model BF6).

Capillaries or ampoules can be insulated in a variety of other ways to give various cooling rates when placed in a refrigerator or plunged into a liquid cryogen. Controlled cooling rates can also be obtained using methanol in a dewar vessel. The methanol is agitated continuously using a mechanical stirrer and the dewar is placed inside a larger dewar containing liquid nitrogen or ethanol–solid CO_2. The cooling rate can be varied by altering the amount of methanol in the inner dewar and by substituting double-silvered or unevacuated dewars. The samples are suspended in the methanol. This apparatus and the cooling rates that can be produced are described for *Leishmania* by Callow and Farrant (1973).

B. Rapid cooling

Since human pathogens have to be sealed in capillaries or ampoules, the fastest cooling rates are those achieved by immersing the sample container directly in a cold liquid. Liquid nitrogen at its boiling point ($-196°C$) is most used, but here a gas envelope forms around the sample container, restricting the rate at which heat can be withdrawn from the sample. With capillary tubes, the fastest cooling rate that can be achieved by direct plunge into liquid nitrogen is around 3000–5000°C per minute, depending on the size of capillary. The insulating layer of gaseous nitrogen is considerably thinner if the sample is instead plunged into nitrogen at its melting point ($-210°C$), and this will almost double the cooling rate attainable compared with nitrogen at boiling point.

Melting-point nitrogen can be made in the laboratory by reducing the barometric pressure which causes a depression of the boiling point. Nitrogen boils off faster under this reduced pressure and the remaining liquid is cooled by the loss of latent heat of evaporation. The liquid remains at the lowered temperature during the return to normal pressure. By removing sufficient heat from the nitrogen in this way it can be converted to a white solid at $-210°C$. The simplest method of producing melting-point nitrogen is in a standard freeze-drying apparatus, but this procedure is potentially hazardous and expert advice should be sought before attempting this.

Rapid cooling rates can also be achieved by plunging the sample container into a very cold liquid, such as isopentane, pre-cooled to −160°C. Heat loss from the sample is extremely rapid since no insulating gas envelope is formed. For non-pathogenic parasites, the sterility of which is not important, even faster cooling rates can be achieved by plunging small sample volumes spread on glass or mica slivvers into liquid nitrogen or another cold liquid, or by introducing the sample in droplet or spray form into the cold liquid.

C. Stepped cooling

For laboratories with access to a low-temperature refrigerator/freezer, the temperature of which can be regulated, this is one of the simplest methods of cryopreservation. The refrigerator is preset to the desired temperature, usually around −25 to −35°C, and the sample simply placed in the cooling chamber for the required length of time and then rapidly transferred directly to the storage temperature.

For *Entamoeba* a holding period of 20 min at −25°C was found by Farri *et al.* (1983) to give good survival, while for *Plasmodium* the optimum holding period was determined by Wilson *et al.* (1977) to be 30 min at −32°C. Ampoules containing the parasites were brought rapidly to the intermediate temperature by placing them in a methanol bath in a precooled refrigerator. Cunningham *et al.* (1964) also reported that good survival of *Trypanosoma brucei* could be obtained with a two-step cooling method in which the samples in capillary tubes were held at −30°C for between 3 and 5 min.

D. Storage and documentation

Before setting up a bank of cryopreserved parasitic species, some time should be spent in deciding whether the samples will be stored in glass capillaries or ampoules, or in plastic ampoules, and what quantity of these containers is likely to be held at any one time. Other considerations will be the ease of transport of the liquid or vapour-phase nitrogen storage container, the frequency with which samples are to be deposited or removed, the static evaporative loss of nitrogen from the vessel, and the availability of a local liquid nitrogen supply. These and other ergonomic considerations will define the size and type of storage vessel and the method of inventory control—on canes, in goblets, or in trays and racks.

Whichever system is finally used it is imperative that comprehensive up-to-date records are maintained of the material deposited and of the additions to, and removals from, the bank. A card or computer file of each sample should include:

(1) Details of its origin (genus, species, and strain of parasite, species of donor host or details of patient, geographical location, date, site of infection in host, method of isolation and name of person making the isolation).

(2) Details of any laboratory maintenance or animal passage (including the animal species, size of inoculum, duration of each infection or passage and date).

(3) Details of the preservation (including person doing the cryopreservation, date, level of parasitaemia, type and amount of cryoprotectant, container, schedule of cooling, and a note of the preferred warming method to be used).

(4) There should also be a record of any viability assays made with thawed material (e.g. count of motile organisms, morphology, infectivity as percentage development relative to a known standard or time taken to reach patency or 2% parasitaemia or the ID_{63} ml^{-1}, pathogenicity, and date).

(5) If the material was cryopreserved as a large batch of tubes or capillaries then a section should be included for recording when these were thawed, to whom they were issued and the number remaining.

Each glass or plastic ampoule should have a unique identification number or code written clearly and indelibly on the outside and its position in the nitrogen storage vessel entered in the card or computer record for that sample. Labelling of individual capillary tubes is difficult, so these are usually put into holders and relevant information written on the holder. Capillaries can be stored horizontally in trays or, more commonly, vertically in goblets. For this latter method, capillary holders can conveniently be constructed out of triangular section plastic spines (the sort used for loose-leaf binding of documents) (Kimber, pers. comm.). A piece of plastic spine is cut with a small saw so that it is approximately 2 cm longer than the capillaries. Two of the three sides are then shortened to leave a 2 cm long tag or handle on the third side, which is used for labelling. The tubes are prevented from falling out of the bottom of this holder by heat-fusing a triangular piece of nylon gauze to the base. Each holder of this type will accommodate approximately 40 capillaries, and the holders can be arranged within the goblets like segments of an orange.

E. Thawing

Almost without exception, most workers use a relatively rapid warming method–agitation of the sample in a water bath, usually set to 37°C but occasionally 40 or even 45°C. The warming rate produced will depend on the volume of the sample and the type of the material from which the container is constructed. For a 20 μl sample in a glass capillary tube, the warming rate will be in the region of 8000°C per minute, while for 1 ml in a polypropylene ampoule the rate will be about 200°C per minute when agitated in a 37°C water bath. For samples cooled slowly or at their optimum cooling rate, the warming rate will not greatly affect the level of survival. However, if the cooling rate was fast, then the optimum warming rate will also be fast and survival will be much reduced at slow rates of warming.

While there has been very little experimentation to determine the effect of cooling rate on survival of particular parasites, there has been even less testing of the effect of warming rate. Only with trypanosomes (Cunningham et al., 1964) and Plasmodium (Mutetwa, 1983; Mutetwa and James, 1984a,b) have different warming rates been evaluated methodically. It was found that rates between about 8 and 2000°C per minute were optimal for trypanosomes previously cooled at 1°C per minute to −60°C and plunged into liquid nitrogen. For Plasmodium cooled at 3600°C per minute, the optimum warming rate was around 8000°C per minute, and this could be achieved simply by removing the capillary tube from liquid nitrogen and agitating it in a water bath at 40°C.

The finding of Whittingham et al. (1979) that mouse embryos cooled slowly (0.3 to 0.57°C min^{-1}) to −80°C survive very much better following slow (20°C min^{-1}) rather than rapid (500°C per minute) warming, suggests that the levels of survival obtained so far with many species of cryopreserved parasites could be improved if investigations were made into variations of the warming rate.

It cannot be said too frequently that improperly sealed tubes or ampoules, into which liquid nitrogen has penetrated during storage, pose a significant hazard during thawing. These will explode and should therefore not be warmed directly in a water bath, but in a beaker or measuring cylinder containing pre-warmed water. If pathogens are liberated by an exploding container, then the organisms will be restricted to the beaker or measuring cylinder, which can then be appropriately disinfected.

IV. SURVIVAL ASSAYS

Parasitologists have in general been particularly bad at quantifying the survival of their cryopreserved organisms. This is not a particularly important issue when material is simply being banked according to pre-determined optimal schedules. However, since in many instances the optimum parameters for cryopreservation have still to be defined, experimentation with techniques will form at least some part of the operations of most workers. Even in the context of experimentation, many published reports have not given the numbers of parasites used, or the volume and concentration either of the material before cryopreservation or upon inoculation into a recipient host or into culture. Furthermore, many workers have failed to use quantitative viability assays for thawed material and have not related the duration of a 'pre-patent period' to any controls or to a reference standard.

Anyone beginning experimental cryopreservation studies with parasitic protozoa should give serious consideration to the available methods used for assaying post-thaw viability.

ANIMAL CELLS

8. MAINTENANCE OF ANIMAL CELLS

A. DOYLE and C. B. MORRIS

European Collection of Animal Cell Cultures,
PHLS Centre for Applied Microbiology Research,
Porton Down, Salisbury, UK

I. Introduction 227
II. Categories of animal cell cultures 228
    A. Primary cell cultures 228
    B. Finite cell lines 228
    C. Continuous cell lines 228
    D. Transformed cell lines 228
    E. Hybridomas 229
III. Growth media 229
IV. Facilities for handling cell lines 230
V. Subculture procedures 232
    A. Adherent cell lines 232
    B. Suspension cell lines 234
    C. Viable cell counts 234
VI. Cryopreservation 236
VII. Quality control 239
    A. Species and cell function verification 239
    B. Detection of contamination 239

I. INTRODUCTION

Animal cell culture emerged as a valuable research tool in the 1940s and 1950s, with classic work in the development of cell culture media (Eagle, 1940; Earle et al., 1943) as one of the preliminary steps. Today, many aspects of research and development involve the use of animal cells as in vitro model systems, substrates for viruses and in the production of diagnostic and therapeutic products (Griffiths, 1985; Montagnon, 1989). Indefinite maintenance of cell lines in culture is not only impractical but would carry the inherent risks of genetic drift and microbial contamination. Techniques have advanced to permit almost unlimited storage of cell lines at liquid nitrogen temperatures

MAINTENANCE OF MICROORGANISMS 2nd Edn
ISBN 012–410351–0

(−196°C) if correct cryopreservation techniques are followed (Doyle *et al.*, 1989a). Many well-characterized, established cell lines are available from commercial suppliers and culture collections (Doyle *et al.*, 1989b).

II. CATEGORIES OF ANIMAL CELL CULTURES

A series of broad definitions can be made concerning categories of cell types available in culture.

A. Primary cell cultures

Cultures isolated directly from animal tissue are known as primary cultures. They can be prepared by disaggregation techniques involving manual and/or enzymatic digestion of adult or, more usually, fetal tissue. They are generally representative of the original tissue with a mixture of cell types present. Such cell cultures can be occasionally propagated onwards ('passage') to produce a cell line.

B. Finite cell lines

Primary cell cultures derived from normal fetal tissue may be passaged for a finite number of population doublings before senescence occurs. With human diploid cell strains (e.g. MRC-5) this can be as high as 70 population doublings (Wood and Minor, 1990); usually, however, with either adult-derived or differentiated cell types, the number of doublings is extremely limited even if material can be established *in vitro* in the first place.

C. Continuous cell lines

Following a 'crisis' phase in a primary or finite cell-line culture (during which time cell growth apparently ceases), a population of cells can emerge with the following general characteristics: reduced cell size, higher growth rate, higher cloning efficiency, increased tumorigenicity and variable chromosome complement. This often results from spontaneous transformation *in vitro*.

D. Transformed cell lines

Normal cells are capable of transformation to a continuous cell line with the characteristics outlined above. However, transformed cell lines can also be derived directly from tumour tissue and also by *in vitro* transformation of cells using whole virus (e.g. SV40, EBV) or DNA fragments derived from a transforming virus using vector systems. It is always difficult to generate transformed cell lines with highly differentiated characteristics, and in long-term culture there is continuous selection for a cell type with the highest growth rate and hence showing the least functional differentiation.

E. Hybridomas

Monoclonal antibody secreting hybridomas are produced as a result of fusion of antibody-secreting B cells from an immunized patient or animal with a malignant myeloma cell line capable of indefinite growth *in vitro*. The cells are cloned and selected for stability. However, this cell type is inherently unstable and can be difficult to maintain; this is especially true if the two fusion partners are derived from different species (e.g. mouse × human hybridomas).

III. GROWTH MEDIA

Cell-culture media aims to mimic the physiological conditions within tissues. This has a near neutral pH, and normally involves the incorporation of serum at concentrations varying from 5 up to 20%, although certain production processes and experimental procedures require the use of serum-free conditions (Maurer, 1986). The most common source of serum is bovine; this may be of adult, newborn or fetal origin.

In addition, there is a complex defined component of the growth medium comprising a buffer system (phosphate, bicarbonate, CO_2 or Hepes), amino acids (including glutamine), vitamins and glucose. The choice of a particular type of medium depends upon the cell type, for example hybridomas are usually cultured in DMEM (Dulbecco's Modified Eagles Medium), and human lymphoblastoid cell lines are cultured in RPMI 1640 (Roswell Park Memorial Institute 1640).

All the components required for tissue culture media are available from commercial suppliers (see Appendix). This can be obtained as powdered medium that requires reconstitution in deionized or distilled water and filter sterilization, or alternatively in 1×, 5× or 10× concentration in liquid form. Care has to be taken in the selection of serum, and suppliers are willing to provide samples for batch testing before a major purchase. The procedures for testing depend upon the techniques in use.

Testing can be as straightforward as the passage of human diploid cells in the test serum, with examination of cells for morphological signs of toxicity and determination of plating efficiency or a more elaborate cloning efficiency study using a hybridoma or tumour cell line (Freshney, 1987).

IV. FACILITIES FOR HANDLING CELL LINES

Traditionally, cell culture facilities have ranged from the minimum, i.e. working on an open bench, up to a specifically designed suite of rooms built to clean room standards. In fact, the routine use of antibiotics such as penicillin and streptomycin in the medium frequently relates to a lack of properly ventilated cell-line handling facilities, where contamination risks are at their highest.

To comply with current safety regulations, a cell-culture laboratory should be fully ventilated, preferably with Hepa filters on the inlets, and equipped with Hepa-filtered workstations where the air flow is directed away from the operator, for example Class II containment cabinets (Fig. 1).

Ideally, when designing cell culture facilities, they should comply with the requirements for Clean Rooms used in industry. To comply, equipment must be kept to the minimum required for the job, there must be proper entry facilities and internal surfaces must be easy to clean and dust free. Such facilities are an expensive but worthwhile investment.

A separate cabinet or room for each cell line handled would be the perfect situation. However, using separate rooms for the different types of operation is a more practical approach. Preparation of reagents and media should not be carried out in rooms where cell cultures are handled. Also, a room designated for receiving or starting new cultures should be available to provide isolation prior to testing cultures for microbial contaminants, especially mycoplasma. Only when cell lines have been screened should they be handled in another room for routine culturing. All normal operations should be feasible without the routine use of antibiotics.

Good management of the facilities requires that certain standard procedures are followed.

(1) Only designated personnel should enter each area and they must be wearing laboratory coats or sterile clothing, including gloves. It may be considered necessary to wear shoe covers, masks and hats.

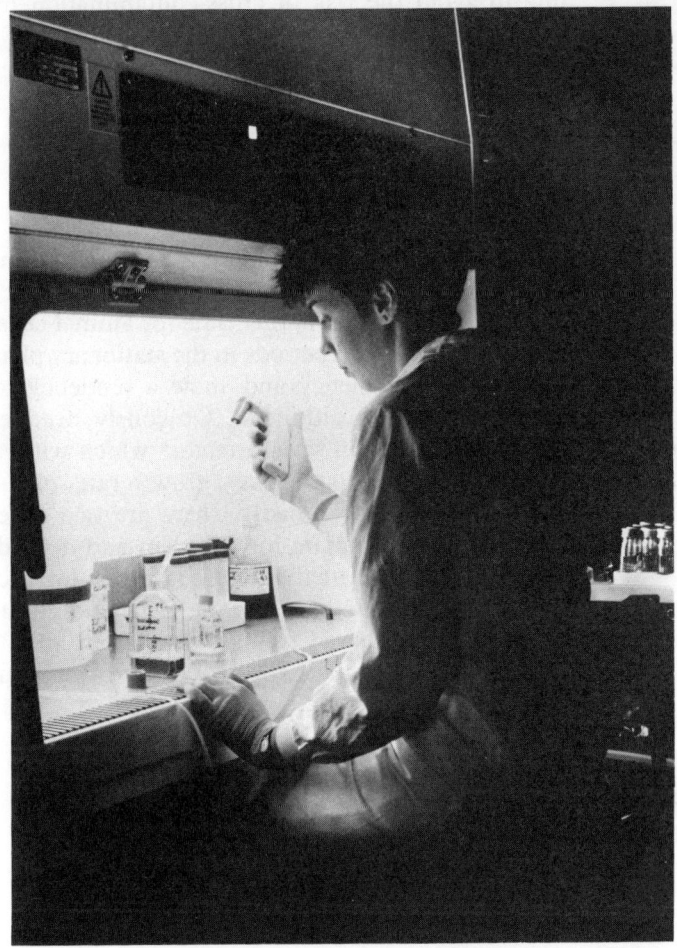

Fig. 1   A Class II containment cabinet for cells. Note the operator wearing protective clothing.

(2) Only sterile, wrapped items (i.e. pipettes, culture flasks, etc.) should enter the rooms, and discarded media, etc., should be removed each day. Germicidal solutions (e.g. 2%v/v Tegodor, Goldschmidt Ltd) should be available for discarding pipettes.

(3) No more than one cell line should be handled at a time in each work area. Between the handling of different lines, a 'clearance time' of 10–15 minutes should be allowed and work surfaces should be sprayed with a germicidal solution.

(4) Water sources, i.e. water baths and sinks, should be kept to a minimum. Water baths should be cleaned at least twice a week.

(5) Eating, drinking, smoking and application of cosmetics must not be allowed in any of the working areas.

(6) Ideally, separate bottles of media should be used for each cell line to avoid the risk of cross-contamination by cells from another culture.

V. SUBCULTURE PROCEDURES

In order to maintain cell cultures in optimum conditions it is essential to keep cells in the log phase of growth as far as is practicable. The usual doubling time for animal cells is 24–46 h. Cells kept for prolonged periods in the stationary phase of growth will lose plating efficiency and show a tendency to decline in overall quality and 'healthiness'. Obviously, frequency of subculture is dependent on several factors which will vary between cell lines, viz. inoculation density, growth rate, plating efficiency and saturation density. Broadly, there are two general types of culture method appropriate for adherent and non-adherent cells and a general outline methodology is given for each. Cultures may be supplied as either growing in tissue culture flasks or frozen in a cryopreserved state. If the latter, then information is given below on resuscitation of cells from liquid nitrogen storage.

Materials:
Phosphate-buffered saline (PBS)
Trypsin–EDTA solution
Growth medium
Tissue culture flasks/Techne stirrers
Graduated pipettes
Centrifuge tubes
Inverted phase-contrast microscope
Centrifuge

The general principles applicable to adherent and non-adherent cell types are given below.

A. Adherent cells

Cells are examined routinely using an inverted microscope at ×100 magnification, and once confluency is achieved (i.e. the cell sheet is complete) the cells should be subcultured to maintain growth. This requires removal of the culture medium, washing the cell sheet with PBS, removing the cell sheet into suspension using a proteolytic enzyme (usually trypsin) and dispensing cells into new tissue culture flasks. The precise method is given below;

it must be emphasised that work must be conducted under strictly aseptic, sterile conditions. Cells may be cultured in tissue culture flasks of 25–175 cm^3, depending upon the yield of cells required. Other methods are available for more large-scale culture of adherent cells in suspension on microcarriers, if large volumes of cells are required (Griffiths, 1986).

(1) Discard spent culture medium.

(2) Gently pour or pipette a volume of PBS half that of the original culture medium.

(3) Swirl the PBS around the flask taking care not to perform this action too vigorously. Discard the PBS.

(4) Add sufficient trypsin–EDTA solution to just cover the cell sheet, i.e. 1–2 ml per 25 cm^2. Allow it to stand on the cell sheet for 30–45 seconds at room temperature and pour off most of the trypsin–EDTA, leaving behind a small volume (c. 0.2 ml). Incubate at 37°C and check for cell detachment after 5–10 min. Vigorous (but careful) agitation of the flasks may aid cell detachment. Some cell types may take longer to remove.

(5) Resuspend cells in a suitable volume of medium; the volume is dependent on the size of the culture vessel but a minimum is required to neutralize the activity of trypsin. If cells show a tendency to clump, it may be necessary to centrifuge the cell suspension and wash in either culture medium or PBS. Such tasks should be carried out with the minimum centrifugal force required to pellet the cells and also avoiding excessive agitation during resuspension, otherwise, considerable damage will be incurred by the cells.

(6) Disperse the cells to produce a suspension and perform a viable cell count (see below).

(7) A cell count may not always be necessary if the cell line has a known split ratio, for example 1:2, 1:4 or 1:16, and the appropriate dilution factor can be used.

(8) Prepare tissue culture flasks with culture medium pre-warmed to 37°C. The volume of culture medium required is 5–7 ml for a 25 cm^2 flask, 20–30 ml for a 75 cm^2 flask and 50–100 ml for a 175 cm^2 flask.

(9) Gas with a 5% CO_2 and 95% air mixture through an on-line 0.2 μm filter for 30–60 s.

(10) Add the appropriate volume of cells to seed the flask. The size of culture flask used depends entirely upon the number of cells required. Saturation densities for adherent cell lines range from about 10^5 cells cm^{-2} for monolayers to 3 to 4 × 10^5 cells cm^{-2} for cells forming multilayers.

(11) Incubate at 37°C.

B. Suspension cell lines

As for adherent cells, cultures are examined microscopically for signs of cell deterioration, i.e. lysis or death, which are signs of overgrowth. Additionally, the colour of the medium can be used as an indicator of cell density: red-orange is normal, whereas yellow indicates that the culture has become too acidic and is at risk of going into stationary phase and decline.

(1) Take a sample for the culture and perform a viable cell count (see below).
(2) Prepare tissue culture flasks or Techne stirrer flasks with culture medium pre-warmed to 37°C.
(3) Gas with a 5% and 95% air mixture through an on-line 0.2 μm filter for 30–60 s.
(4) Add the appropriate volume of cells to the new flask. Incubate at 37°C. Normally, the viability of suspension cells should not be allowed to fall below 90%. However, cultures of lower viability may be recovered by subculture, although it will be necessary to pellet the cells by centrifugation and decant the spent medium.

A typical growing density range for many suspension cells is 10^5–10^6 cells ml^{-1}, although some lines may require higher dilution or may reach higher saturation densities.

C. Viable cell counts

Both during subculturing and cryopreservation procedures it is important to estimate the percentage of viable cells in a population. The simplest method is to examine the cells microscopically in a counting chamber (haemocytometer) using a vital stain, trypan blue, although other methods are available (Baserga, 1989).

Materials:
Haemocytometer (improved Neubauer type)
0.4% trypan blue (w/v) in PBS
Culture medium
Tally counter
Pasteur pipettes
Graduated pipettes
Phase-contrast microscope (×100 magnification)

Procedure:
(1) The haemocytometer (Fig. 2) must be thoroughly clean before use. It can be left to soak in 70% alcohol or 0.1 N HCl to remove proteinaceous material.

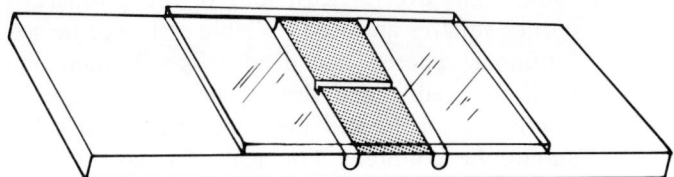

Fig. 2   A counting chamber (haemocytometer) with the coverslip in place.

(2) Wash the haemocytometer under running water and dry. Place a clean coverslip (a special type, supplied with the haemocytometer, is required) over the grid and press down gently at the edges to seal the glass surfaces. The appearance of Newton's rings indicates this has been achieved.

(3) Take an exact aliquot of cell suspension (50–100 μl) and mix with a known volume of trypan blue solution. The amount will depend on the cell concentration; normally, a dilution of between 1:1 and 1:10 will provide sufficient cells to count.

(4) Using a Pasteur or micro-pipette, fill one side of the chamber by allowing the suspension to be drawn under the coverslip by capillary action.

(5) Examine the marked grid under a ×10 objective (×100 magnification). The grid is divided by lines and when examined as a whole is subdivided into nine large squares, each of which is further subdivided. The area of each of these large squares is 1 mm² (Fig. 3).

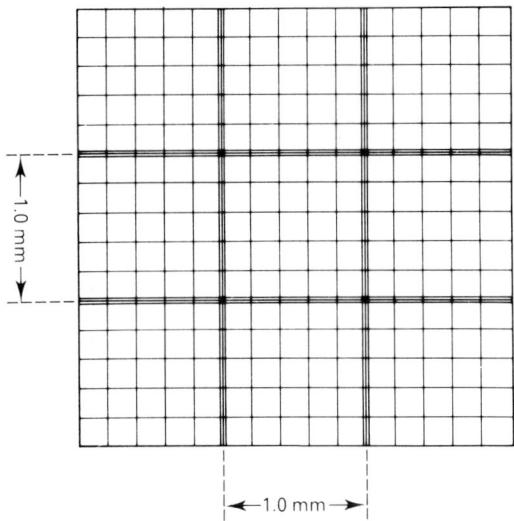

Fig. 3   The counting grid, showing its division into nine 1 mm² squares.

(6) Count cells over at least four of these squares (usually the corner squares are used). Viable cells will be seen as bright, luminous spheres, and dead cells will stain blue. Calculate both the viable and dead cells. For acceptable accuracy, not less than 30 and not more than 200 cells per 1 mm² should be counted. The use of a tally counter improves accuracy. Adjust the ratio of cell suspension to trypan blue if necessary.

(7) Cell counts are calculated as follows:

$$\text{cells ml}^{-1} = \frac{\text{No. cells counted}}{\text{No. 1 mm}^2 \text{ squares counted}} \times \text{dilution} \times 10^4$$

To calculate the viability of a cell suspension, divide the number of viable cells by the total number of cells (i.e. viable + dead). Healthy cultures should be more than 90% viable. A viability less than this indicates that either the cells are being exposed to sub-optimal growth conditions or they have exausted their nutrient supply.

VI. CRYOPRESERVATION

It is impractical for most laboratories to maintain cell lines in culture indefinitely; moreover, cell cultures will undergo genetic drift with continuous passage, and risk losing their differentiated characteristics. Therefore, it becomes necessary to store cell stocks for future use. It is recommended practice that a limit is set for the number of passages any cell line should undergo before replacing from cryopreserved stocks. This is of less importance for undifferentiated cell lines having infinite lifespan, although it still remains a consideration.

Nearly all cell lines can be cryopreserved successfully in liquid nitrogen at $-196°C$. A few simple criteria must be adhered to in order to provide sufficient cells for all future needs.

The essentials for efficient cryopreservation are slow freezing, i.e. at a rate between 1 and 3°C per minute, and fast thawing, which is achieved by placing ampoules in a water bath at the temperature required for growth. The addition of a cryoprotectant, such as glycerol or dimethyl sulphoxide (DMSO) enhances survival of cryopreservation. Nevertheless, DMSO does have certain toxic properties. Therefore care must be taken in handling the cryoprotectant, and some consideration may be necessary as to whether cells should be washed immediately on thawing.

A full discussion of problems encountered in cryopreservation is available elsewhere (Doyle et al., 1989a).

Materials:

Sterile glass or plastic ampoules (1.5–2.0 ml capacity; e.g. Wheaton, glass, or Nunc, plastic).

Sterile pipettes or automatic dispensing apparatus

Racks to hold the ampoules (Nunc)

Improved Neubauer haemocytometers

0.4% (v/v) Trypan blue in PBS

Freezing solution, i.e. growth medium containing 20–25% serum or complete serum (e.g. newborn calf serum) and 7–10% cryoprotectant (high grade glycerol or DMSO; newest stock available from supplier)

Protective face mask and gloves for handling liquid nitrogen

A programmable freezer (Planer) or two-stage freezer

Liquid nitrogen storage vessel (Union Carbide, Cryo-Med)

The decision on whether to use glass or plastic ampoules will depend upon whether hermetically sealed ampoules are required. Complete enclosure, eliminating nitrogen, can only be achieved with sealed glass ampoules. However, this requires expensive equipment if large numbers are to be produced, and pre-testing for leaks is essential to avoid sudden explosions on thawing of ampoules containing nitrogen. For most laboratories it is more practical to use the screw-cap plastic ampoule. These are made of a special grade of plastic designed to withstand extreme cold (cryotubes). Even though they are more convenient to use, precautions are still necessary in handling ampoules of this type (see below).

Method:

(1) Only freeze cultures that are healthy and in log phase growth.

(2) Perform a viable cell count as outlined in Section 3 above. If the cultures contain more than 20% dead cells it is advisable not to freeze them. Cells can be fed by a change of medium, and subsequently examined to see whether viability has increased.

(3) Centrifuge sufficient cells to fill the required ampoules with at least 4×10^6 cells per ampoule. The maximum cell number permitted should not exceed 2×10^7 cells per ampoule.

(4) Prepare sufficient freezing medium to fill the ampoules with 1 ml each. A recommended medium is whole serum (e.g. newborn calf (or fetal bovine) serum to which 7–10% (v/v) cryoprotectant is added. By eliminating culture medium, which usually contains bicarbonate, a pH closer to neutral can be maintained during the freezing process. It is possible to use 20% serum with culture medium and cryoprotectant,

but great care is necessary to avoid a change to alkaline pH during dispensing of ampoules.

(5) Resuspend the cell pellet in the freezing medium and dispense into the ampoules in 1 ml aliquots. Clearly mark the ampoules with the cell designation, passage number and date of freezing.

(6) Cool the ampoules at between −1 to 5°C per minute until at least −60°C is reached. The most suitable rate can be determined by prior experimentation. At the European Collection of Animal Cell Cultures (ECACC), −3°C per minute is used for most cell types. A programmable freezer can achieve this by means of an electronically controlled input value on the nitrogen supply to a cooling chamber. When such equipment is not easily available to the researcher, an alternative method involves freezing cells in a −70°C freezer using an insulated container. Ampoules are placed in a polystyrene box or block so that each ampoule is insulated by about 1.0–1.5 cm of material, including the top and bottom of the ampoule. Place the box in a central position at a quarter to a third from the top of the freezer. Leave the ampoules to cool for 16–24 h, then immediately transfer them to a nitrogen storage vessel. If animal cells are kept at −70°C for more than a few months their viability will rapidly deteriorate.

A simpler method is to use the two-stage freezer where ampoules are placed at a pre-determined depth in the neck of a dewar flask containing liquid nitrogen. After a period of 15–20 min the ampoules can be plunged directly into liquid nitrogen. Further discussion of cryopreservation is available in Ashwood-Smith and Farrant (1980).

(7) After freezing, the ampoules are immediately transferred to a nitrogen storage vessel and kept in either the liquid or gas phase.

(8) Ampoules for resuscitation should be thawed rapidly. Place them in a water bath at the temperature required for growth, for example 25°C for amphibian cells, 37°C for mammalian cells. It is advisable to place ampoules inside a screw-capped metal container with holes to allow free passage of water. This device offers protection if the ampoule should explode. To avoid possible explosion from trapped nitrogen, allow the ampoule to stand at room temperature for 2 min before placing in the water bath. During all stages of handling the ampoule, full face and hand protection should be worn.

(9) Once thawed, the ampoule should be wiped with a tissue soaked with ethanol. The contents are transferred with a pipette to a 15 ml centrifuge tube. Slowly add 0.5–1.0 ml of growth

medium to the tube and either transfer it to a prepared culture flask or centrifuge the cells to remove the DMSO. Centrifuging is only required if it is considered necessary to remove the cryoprotectant.

(10) It is advisable to start cultures at between 30%–50% of their final (maximum) cell density. This allows the cells rapidly to condition the medium and enter into the log phase of growth.

Maintenance of accurate records of stored ampoules is very important. In addition to written records it is now possible to purchase inexpensive computer software for storage documentation, for example Cardbox Plus from Business Simulations Ltd, and maintain records on a personal computer.

There are several manufacturers of liquid nitrogen storage vessels (Union Carbide, Cryo-Med). Tanks are available with capacities ranging from as few as 32 ampoules to over 25 000 ampoules capacity (Fig. 4). There are also electrical freezers operating at $-135°C$. It is important in all cases to have contingency plans for failure of integrity of liquid storage vessels, or electrical failure with $-135°C$ freezers. This requires having alarm systems attached to the vessels. Also, it is essential to maintain valuable cell stocks in more than one location, preferably using safe deposit facilities in a culture collection for irreplacable material.

VII. QUALITY CONTROL

A. Species and cell function verification

For each cell type cultured there will be appropriate techniques available for the verification and authentication of the cells. These could depend upon functional assays (e.g. antibody production) or verification of species by cytogenetic or isoenzyme analysis. Further information is given by Doyle et al. (1989b, 1990).

B. Detection of contamination

As already stated, aseptic conditions are essential for all cell culture work. Contamination by bacteria or fungi is normally easily apparent to the naked eye, and contaminated cultures should be discarded. However, there is one problem area of contamination of cell lines — mycoplasmas. These are the

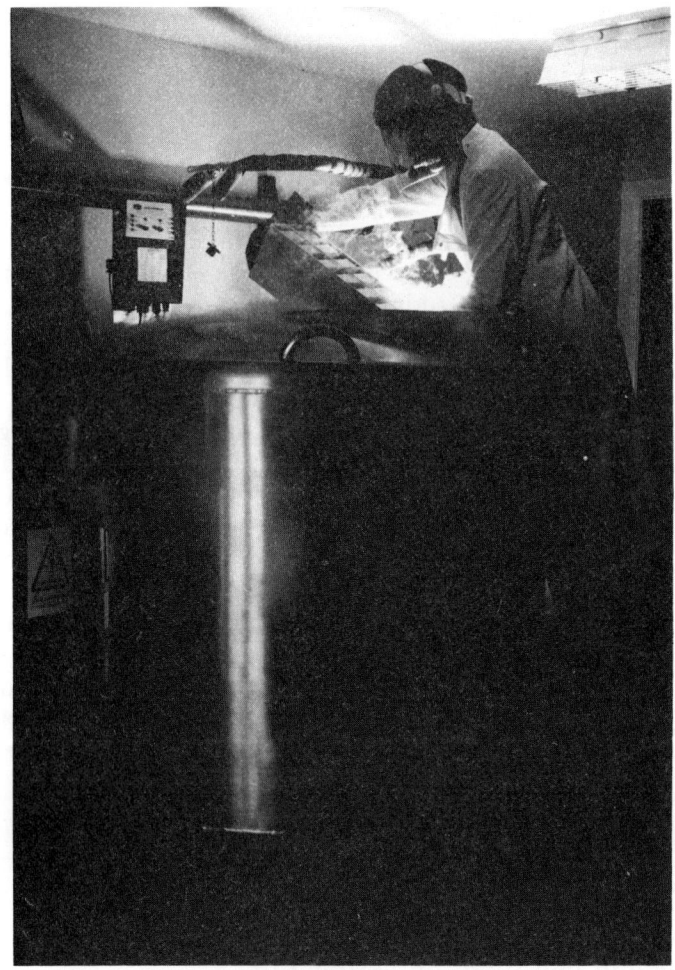

Fig. 4 Liquid nitrogen refrigerator with a capacity of approximately 25 000 ampoules (ECACC cell storage laboratory).

smallest self-replicating organisms and can be as little as 0.3 μm in diameter. Therefore they are not readily apparent and special techniques are required for their detection. Several test procedures can be used, including microbiological culture, Hoechst 33258 DNA stain together with fluorescence microscopy, indirect toxicity assays using 6,6-methylpiredeoxyriboside, and DNA probes (McGarrity and Kotani, 1986; McGarrity *et al.*, 1986). The most readily accessible technique is the Hoechst 33258 DNA stain and, because of the fundamental requirement for mycoplasma-free cultures in tissue culture, an outline of the technique is given here.

Hoechst 33258 stain for mycoplasma detection:

(1) Before testing, cells should have been cultured, antibiotic-free, for a minimum of two passages.

(2) Harvest adherent cells by the usual subculture method and resuspend cells in the original culture medium at $c.\ 5 \times 10^5$ ml^{-1}. Suspension cells can be tested directly at a similar concentration. Add $c.$ 1 to 5×10^4 cells to each of two 22 mm coverslips in 35 mm diameter tissue culture dishes. Incubate cells overnight at 37°C in a humidified atmosphere of 5% CO_2 and 95% air. Test one sample at 24 h, and examine the remainder after 72 h.

(3) Before fixing, examine the cells on an inverted microscope ($\times 10$ magnification) for evidence of microbial contamination.

(4) Add 2 ml Carnoy's fixative (methanol–acetic acid) dropwise from the edge of the culture dish and leave for 3 min. Care must be taken not to sweep cells to one side of the dish. Pour the fixative into a waste bottle and add another 2 ml of fixative. Leave for a further 3 min.

(5) Dry the coverslip in air on the inverted tissue culture dish lid for 30 min. Wearing gloves for handling, add 2 ml freshly prepared Hoechst 33258 stain (0.05 μg ml^{-1}) and leave for 5 min. Shield the coverslip from light at this point. Decant the stain to a waste bottle.

(6) Add one drop of mountant to a glass slide and mount the coverslip cell side down.

(7) Examine under ultraviolet fluorescence at $\times 100$ magnification. Uncontaminated cells show only brightly fluorescing cell nuclei, mycoplasma appear as cocci or filaments in the cytoplasmic area. Positive control slides do assist in detecting contamination. Contaminated cultures should be discarded. Information on elimination of contamination is available elsewhere (Mowles, 1988).

Under routine cell culture conditions, cells should be routinely examined for mycoplasmas every 4–6 weeks. Quality control is not a single one-off exercise but a continuous monitoring process and, in the long run, can prevent a number of major problems.

PLANT TISSUE CULTURES

9. MAINTENANCE OF PLANT TISSUE CULTURES

LYNDSEY A. WITHERS

International Board for Plant Genetic Resources,
c/o FAO, Rome, Italy

I. Introduction 243
  A. Culture systems and their applications 243
  B. Culture storage — background 244
II. Broad approaches to culture maintenance 246
  A. Advantages and disadvantages of normal growth
    storage 246
  B. Slow growth............................... 247
  C. Suspension of growth 248
III. Slow growth................................... 248
  A. Callus cultures — a question of balancing risks ... 248
  B. Shoot cultures — practical applications 249
IV. Cryopreservation 250
  A. Principles 250
  B. Cell suspension cultures 253
  C. Protoplasts 257
  D. Callus 259
  E. Shoot cultures 259
  F. Somatic embryos 262
  G. Cryopreservation equipment.................. 264
V. Continuing problems and research challenges in
  cryopreservation 265
  A. Stability in cryopreserved cultures 266
VI. *In vitro* conservation in practice................... 266

I. INTRODUCTION

A. Culture systems and their applications

Plant tissue culture, otherwise known as *in vitro* plant culture, is a vast area of technology encompassing the study of systems ranging from protoplasts — ephemeral protoplasmic units isolated from the confines of the cell wall — through cell suspensions and

MAINTENANCE OF MICROORGANISMS 2nd Edn
ISBN 012–410351–0

callus cultures to shoots, roots, embryos and plantlets. These systems are very different both in dimensions and structural complexity. They can be subjected to generalization at only the most superficial level, i.e. broad similarities in culture media and physical growth environment. Even then, precise culture conditions vary widely with culture system, genus, species and even variety. Thus, any consideration of maintenance can be built only upon a good knowledge of basic culture procedures. The information given in this chapter is not a substitute for basic knowledge but a supplement to it.

Application of plant tissue culture technology lies in a number of spheres: crop improvement, mass propagation, secondary product synthesis, plant physiology, biochemistry and molecular biology, and plant genetic conservation. All of these activities involve the production and maintenance of cultures of various types; in plant genetic conservation, maintenance is the central objective (Withers, 1989).

B. Culture storage — background

In the history of the development of techniques for the maintenance of plant tissue cultures, there has been a tendency to emphasise unorganized culture systems in activities such as fundamental physiological studies and in secondary product synthesis (Henshaw 1975; Kartha, 1985; Withers, 1980a, 1982, 1986; Withers and Williams, 1982). Conversely, there has been an emphasis on more organized shoot, embryo and plantlet cultures in mass propagation and conservation. However, the target applications of the different culture systems are overlapping as expertise develops in their manipulation. For example, embryogenic cell cultures hold great promise in genetic conservation (Kartha et al., 1988). Thus, the reader is urged to examine the potential of all culture systems in the context of any new application.

The objective of applying special storage conditions may simply be to relieve the workload of culture maintenance, its costs and its risks. In other cases it may be to 'freeze time' and assist an experimental procedure by storing timed samples. Increasingly, however, storage is being used to conserve the genetic integrity of unique plant genotypes. Clonal fidelity, at one time a by-word in plant tissue culture, is now regarded as by no means guaranteed and is, in any case, a relative rather than absolute phenomenon (Scowcroft, 1985). Various factors threaten genetic stability in culture, and the use of special storage techniques is one of the few lines of defence against this, particularly in the more vulnerable culture systems (see Section II.A). Although all users should always bear in mind this motive for culture storage, it is probably in the areas of biotechnology and plant genetic conservation that awareness of this aspect is greatest.

For the laboratory scientist, storage of cultures is an obvious foundation technology underpinning other investigations; storage by new techniques such as cryopreservation is a natural, welcome progression from the use of serial subculture. However, for the worker in plant genetic conservation, itself a conservative discipline, the adoption of storage *in vitro* involves two significant steps, firstly the conversion of free-living plants into cultures and, secondly, the imposition of highly unnatural conditions to control or suspend the growth of those cultures. A level of caution and a wise unwillingness to abandon more conventional approaches are prominent in current attitudes. Although a certain amount of 'evangelism' may be required to implant the idea of *in vitro* conservation, it is unlikely that there will ever be a complete transition to new technologies. A pragmatic balance of old and new, each complementing the strengths of the other is advisable and feasible. Thus, for any crop, there will be a place for the field genebank, for the *in vitro* collection maintained under conditions of normal and slow growth, for cryopreservation, and for the storage of seed, pollen and, in due course, even DNA (see Withers, 1989, 1991a,b).

Plant tissue culture storage methodology has been under development for some 15–20 years. During that time, considerable progress has been made, particularly in the cryopreservation of living organisms in general and of plant material in particular (Ashwood-Smith and Farrant, 1980; Kartha, 1985; Grout and Morris, 1987). This has happened in parallel with dramatic developments in the field of plant tissue culture itself. Nevertheless, we are not in a position today to offer routine storage procedures for all types of culture of all species. In this chapter, an attempt is made to present the methodologies which are at or nearing a routinely applicable stage of development. Continuing difficulties and deficiencies that must be resolved by research are discussed and some potentially fruitful lines of investigation suggested (see Section V). In so doing, a selection will be made from the relatively large literature on the subject. The wider literature can be accessed through the reference lists of other reviews of the subject, including Grout (1990), Kartha (1985), Sakai (1985) and Withers (1980a, 1982, 1984, 1985a,b, 1986, 1987a,b, 1990).

II. BROAD APPROACHES TO CULTURE MAINTENANCE

A. Advantages and disadvantages of normal growth storage

Transferring plant germplasm to *in vitro* culture is in itself a maintenance procedure but only of the most rudimentary kind. Although it provides some security, in the sense that the material is protected from the external environment, culture under normal growth conditions is not satisfactory for other than very short-term conservation. There are a number of reasons for this. Most culture procedures are designed to give a relatively rapid rate of growth and propagation, to give a large bulk of material or a large number of cloned plantlets in a short time. Frequent subculturing is necessary, each operation exposing the culture to risks of accidental introduction of microbial contaminants. Consumable materials (culture media, vessels, disposable instruments, etc.) are costly, and the labour input for subculturing can be a significant proportion of the technical support to a tissue culture laboratory.

Practical considerations and cost are important factors in selecting a culture maintenance procedure, but even more important is the question of genetic stability. This is a complex issue, and we have yet to gain a full understanding of the nature, causes and means of control. The type of instability manifested *in vitro* and generically termed 'somaclonal variation' (Larkin and Scowcroft, 1981; Scowcroft, 1984) ranges from subtle, single base-pair changes in a nucleotide sequence through to ploidy changes and loss of chromosomes. The most 'risky' culture systems from the point of view of stability are protoplasts, cell suspensions and calluses, i.e. unorganized cultures; the least 'risky' are organized cultures.

In the course of designing maintenance protocols, conditions likely to be conducive to somaclonal variation should be avoided. Under some circumstances, susceptible genotypes can also be avoided; in the genetic conservation of a crop, it is not acceptable to make a selection on the basis of low somaclonal variation risks, and agronomic factors should dominate. Similarly, in the establishment of a collection of cultures of secondary product synthesizing genotypes, the scientist will need to recognize the reality of somaclonal variation and use that awareness to avoid its worst consequences.

Our knowledge of somaclonal variation is increasing steadily and it seems likely that factors such as the level of methylation of DNA are involved (Phillips *et al.*, 1990). The molecular biologists may well have a solution to the problem in the not too distant future. In the meantime, it remains one of the realities of culture initiation and maintenance, and forms part of the balance sheet of *in vitro* conservation.

An example of somaclonal variation that must be confronted is in the case of *Musa* spp. (banana and plantain). This genus includes the familiar dessert banana and the important staple crop, the cooking banana or plantain, plus other types used for brewing or fibre production. The *in vitro* propagation system used for multiplying *Musa*, many genotypes of which are sterile, is neither non-adventitious nor strictly adventitious. Some genotypes show extremely high frequencies of instability (up to *c.* 60%) others show virtually none (Vuylsteke *et al.*, 1988; D. Vuylsteke and Swennen, 1990.). The patterns are consistent in indicating a strong genetic component. The types of variant observed are not unique to *in vitro* culture but are amplified from the frequencies observed in the field. Somaclonal variation in *Musa* will sensibly be tackled not by avoiding *in vitro* procedures but by using and monitoring them judiciously, by avoiding known hazards, including the obvious one of selection under stress, by developing maintenance procedures that reduce risks, and by also developing screening procedures that detect variants at an early stage. The already serious consequences of somaclonal variation are made more serious by their late detection at a stage when a large investment has been made, and a system of early-stage markers is needed to avoid such problems.

It is clear that growth under normal conditions does not meet culture maintenance needs in other than the very short term. Alternatives are needed, and these fall into two categories: slow growth and the suspension of growth.

B. Slow growth

The term 'slow growth' is used here rather than the often quoted 'minimal growth' for reasons of accuracy. Some cultures are naturally slow growing and 'minimal growth' implies the achievement of the minimum tolerable rate of growth through precise experimentation, which is rarely carried out. Several approaches to slow growth have been attempted for higher plant cultures. The most successful involves modifying the culture medium and/ or reducing the growth temperature. Less well explored or widely used methods are mineral oil overlay (Caplin, 1959; Agereau *et al.*, 1986; Crane and Hughes, 1990) and modification of the atmosphere to which the culture is exposed (Bridgen and Staby, 1983; Engelmann, 1990a). These are interesting scientifically and may have specific, appropriate applications, but they cannot be recommended for general use.

C. Suspension of growth

Some success has been achieved in the freeze-drying of pollen (Towill, 1985) but this technique has not been applied successfully to other higher plant materials. Undercooling in oil emulsions has been attempted for a limited number of cell suspensions and shoot cultures, but only short-term storage has been reported and there are very limited data on which to build a conservation method for higher plant cultures (Franks *et al.*, 1983), although the system has great value as an analytical tool in biochemistry. Cryopreservation (storage at an ultra-low temperature in liquid nitrogen) is the only satisfactory method for higher plant cultures.

Thus, the technical options for *in vitro* conservation fall into three categories: normal growth, slow growth and cryopreservation, with the real choice other than for the briefest storage period lying between slow growth and cryopreservation. At the present stage of development of the technology there is more than the time-scale implication in choosing between the two as not all culture systems are equally susceptible to the two approaches. To generalize, slow growth is more suitable for organized cultures and cryopreservation has achieved a higher level of reproducibility for unorganized cultures, especially cell suspensions.

III.   SLOW GROWTH

A. Callus cultures — a question of balancing risks

In circumstances when storage is short- to medium-term, slow growth can be a convenient option. However, it is not advisable in the long-term because of risks of selection due to stress imposed on the culture. This caution relates particularly to unorganized cultures. Therefore, application of slow growth must be made in the light of the risks involved and a cost–risk–benefit analysis performed.

There are examples of morphogenic potential in callus cultures being retained in slow growth over a period of time during which it would have been lost under normal growth conditions. However, other arguments against slow growth storage for callus cultures are demonstrated by the work of Hiraoka and Kodama (1984). They have shown that in less than a year, callus of secondary product synthesizing cultures show poor recovery growth and low and/or erratic yields of secondary products. Callus cultures are renowned for losing specific traits such as this in normal growth, particularly high-yielding lines isolated by repeated cloning

(Deus-Neumann and Zenk, 1984). Therefore, it must be asked whether, on present knowledge and in the light of progress in the application of cryopreservation to callus cultures, slow growth can be judged to be a very useful alternative.

B. Shoot cultures — practical applications

The situation is different for shoot cultures, which are inherently more stable and in which the opportunities for monitoring stability through morphology and performance of regenerated plants are greater. The most well-researched and widely applied methods of slow growth storage for shoot cultures involve reducing the temperature at which the cultures are incubated. Some success has also been achieved by modifying the culture medium to inhibit growth; in some cases the two approaches are combined. Generally, it is found advantageous to reduce the light level to which the cultures are exposed in slow growth. Examples are shown in Table I, to illustrate the types of modification successfully applied to seven species in order to achieve slow growth and so extend the subculture interval to a convenient length without unduly endangering viability. These examples will help in the design of storage protocols for other subjects.

An important difference between slow growth and cryopreservation is the nature and extent of cumulative effects. In the case of cryopreservation, the main risks to viability and stability are attached to the stages that occur once and are of finite duration, i.e. from pre-growth to cooling and from warming to recovery. Effects related to the duration of storage are negligible. Where background radiation might cause the accumulation of mutations over long periods of storage, exposure to enhanced levels of radiation can be simulated from a controlled source. In contrast, slow growth can only be tested in 'real time'. This means that knowledge is being gathered only slowly on the cumulative effects of storage.

Some of the most long-term experiments have been carried out on cassava and potato genotypes at the International Agricultural Research Centers of the Centro Internacional de la Papa (CIP) Peru and the Centro Internacional de Agriculture Tropical (CIAT) Colombia. Data gathered for cassava cultures over a period of 8 years suggest that there may be a gradual 'habituation' of some material to slow growth (W. M. Roca, pers. comm.), giving a warning that no assumptions can be made about the reproducibility over time of slow growth. In any case, cultures maintained in slow growth need careful monitoring to detect loss of viability or vigour, contamination and exhaustion of the medium.

As well as investigating the response of cultures in the relatively long-term, there is scope for research into the modification of factors such as the size and gaseous exchange characteristics

Table I: Slow growth procedures for cultures

Species	Storage conditions	Duration (months)	Reference
Ipomoea batatas (Sweet potato)	Reduced temperature (from 27 to 16 or 18°C); addition of mannitol (3%) to medium; reduced light level (from 3000 to 1000 lx)	12–18	Love *et al.* (1987)
Manihot esculenta (Cassava)	Reduced temperature (from 28 to 23°C); reduced light level (from 3000 to 1000 lx)	18–24	IBPGR–CIAT (1990)
Musa spp.	Reduced temperature (from 30 to 15°C); reduced light level (from 3000 to 1–2000 lx)	12–18	Banerjee and De Langhe (1985), Schoofs (1991)
Prunus spp. (5 species)	Reduced temperature (from 18 to 2°C); defoliated shoots; dark incubation	18	Druart (1985)
Pyrus communis (Pear)	Reduced temperature (from 28 to 4°C)	18	Wanas *et al.* (1986)
Sassaurea lappa (Costus/Kuth)	Reduced temperature (from 25 to 5°C); dark incubation	12	Arora and Bhojwani (1989)
Solanum spp. (potato)	Reduced temperature (from 20 to 10°C); addition of N-dimethyl-succinamic acid (50 mg l^{-1}) to medium; reduced light level (from 4000 to 2000 lx)	24	Mix (1985)

of culture vessels to improve the condition of shoot cultures in slow growth for extended periods of time (see Grout, 1990). For culture systems other than shoots and callus, even basic research into slow growth conditions is lacking. On present knowledge, it is not possible to offer reliable guidelines or even outline procedures.

IV. CRYOPRESERVATION

A. Principles

Cryopreservation results in a suspension of metabolic activity by transfer of the specimen to ultra-low temperatures. Conventionally, liquid nitrogen is used as the coolant and results in a temperature of −196°C in the liquid itself or, if material is held in the vapour phase over a body of liquid nitrogen, a storage temperature of approximately −150°C. There are advantages to the latter storage system as it avoids the risk of liquid nitrogen penetrating the containers (e.g. ampoules) in which the specimens are enclosed, which can cause them to explode upon thawing and possibly injure personnel. The use of plastic ampoules rather than glass reduces this risk.

There are other important safety matters to take into account, including the need to protect the face (particularly the eyes) when handling liquid nitrogen and to wear insulated gloves when touching heat conducting materials in contact with liquid nitrogen (forceps, drawers within a refrigerator, etc.). Liquid nitrogen itself should not come into contact with the skin. Containers of liquid nitrogen should be stored in a well-ventilated area to avoid the build-up of nitrogen gas and associated risk of asphyxiation.

The physical and engineering aspects of ultra-low temperature storage are well worked out and more-or-less uniform across the spectrum of living materials held in storage. The biological aspects, however, vary. Even though broad guidelines can be followed, it is necessary to 'tune' procedures for the specimen in question. Few higher plant specimens can survive transfer to ultra-low temperatures without any preparative treatment. Examples of undemanding materials are desiccated 'orthodox' seeds and dormant buds of winter-hardened trees. These examples give pointers to the treatments that must be applied to sensitive materials to render them tolerant of the changes that allow exposure to ultra-low temperatures and the stresses this incurs.

The stages of cryopreservation are identified as: pre-growth, cryoprotection, cooling, storage, warming, post-thaw treatment and recovery. For each type of culture, conditions will vary at each stage, indeed each stage may have differing degrees of importance in achieving success. Therefore, for details, refer to Sections IV.B to F. Common points are outlined here.

Pre-growth provides the opportunity to select a particularly freeze-tolerant stage of the growth cycle of a culture, to increase freeze tolerance by adjusting the culture conditions, or to modify the size and/or structure of a culture by dissection or filtering. In principle, pre-growth is seen as a stage that allows *transient* changes in a culture rather than irreversible ones. A treatment that is highly selective in that it favours the growth of a proportion of a population and introduces the risk of selection, is contrary to the objectives of genetic conservation. There is a necessity to maintain high percentage survival levels at all stages.

Cryoprotection involves the treatment of the culture with

compounds that enhance freeze tolerance. The classical cryoprotectants, glycerol and dimethyl sulphoxide (DMSO), feature
prominently in higher plant cryopreservation, but it is common
to find other compounds such as sugars, sugar alcohols, amino
acids and polymers used in combination with one or both of
these. Some classification into types based on modes of action has
been attempted. It is not, however, particularly helpful to
attempt to construct a recipe on the basis of such a classification;
rather, it is recommended that the guidelines of previous users be
followed, with some empirical adjustments to suit the specimen
in question. Cryoprotectants are usually prepared in culture
medium rather than water. Organized cultures are less demanding
and can survive with the application of DMSO alone, but unorganized cultures normally require a mixture of cryoprotectants.

Cooling is a critical stage in most procedures and the choice of
rate depends upon the mechanism of survival. A slow rate will
induce cellular dehydration ('protective dehydration') and a rapid
rate will involve freezing of intracellular water with the avoidance
of ice damage through vitrification (see Section IV.B) or microscopically small ice formation. Again, these are only broad
guidelines and attention must be paid to the individual culture
system in question. With the accumulation of experience in
cryopreservation of higher plant materials, it is becoming easier
to predict cooling rates, but it is not yet possible to recommend
one without qualification for all culture systems.

Storage conditions have already been in part addressed above.
It should be added that whilst electrically powered refrigerators
running at approximately $-100°C$ can be acceptable in the short-
to medium-term, it is not acceptable to use refrigerators cooled
by solid carbon dioxide ('dry ice') at -70 to $-80°C$, or electrically
cooled ones in this temperature range. Standard domestic or
laboratory deep freezers running at $-20°C$ are totally unsuitable
as they would not even maintain the frozen state in cells.

Conditions during warming are dictated by cooling conditions.
Thus, rapid cooling must always be followed by rapid warming to
avoid ice-recrystallization damage, whereas slow cooling may be
followed by slow warming. However, the latter combination can
be difficult to achieve successfully without a great deal of
empirical experimentation and may not give any great benefit
(but see also Sections IV.B, F). Therefore, rapid warming to
avoid ice-recrystallization damage is usually the best choice.

Post-thaw treatments are intended to stabilize material that has
survived the traumatic transitions to and from the storage
temperature. In general, this involves minimal physical/osmotic
disturbance, although it may be necessary to remove toxic
cryoprotectants.

'Recovery' is the period (days to weeks) after thawing before the culture behaves normally. During the period, the culture may require supportive conditions in order to maintain viability and encourage growth. In the case of organized cultures, the types, combinations and concentration of growth regulators ('hormones') in the culture medium may need to be modified. In general, recent studies have shown an increase in post-thaw viabilities and less necessity for 'nursing' in all types of culture.

The following sections give examples of cryopreservation procedures that have been demonstrated to be effective for each culture system, with information on alternative techniques that can improve the performance of cultures that do not respond well to the model procedure.

B. Cell suspension cultures

The model procedure (Fig. 1) is based on that of Withers and King, 1980; Withers 1985c; Withers, 1990a, for cell suspensions. Many details are relevant to other systems, and the model serves as a good initial approach to 'home in' on a successful procedure.

Model procedure

1. Pre-growth Using knowledge of the growth cycle of the culture, harvest cells in early exponential/rapid growth when they are small and highly cytoplasmic. If the cells are relatively large and vacuolated, use a modified medium containing mannitol at 3 or 6% in the passage before cryopreservation. Chill the suspension on ice.

2. Cryo-protection Prepare a solution of 1 M DMSO, 1 M glycerol and 2 M sucrose in standard culture medium and adjust the pH to the standard value; filter sterilize using pressure/suction if necessary as the mixture is very viscous. Warming slightly will help to reduce the viscosity, but the mixture must not be heated excessively or autoclaved. Chill the mixture on ice. Add to an equal volume of chilled cell suspension. Mix well. Incubate at 4°C for 1 h, stirring continuously or swirling periodically to avoid layering. Dispense into polypropylene ampoules. Seal and label with a suitable marker (i.e. not alcohol-soluble). Etching with a hot needle provides a simple marking method resistant to all cooling and storage conditions.

3. Cooling Cool at 1°C per minute to −35°C. Hold at −35°C for 40 min and then plunge into liquid nitrogen.

4. Storage Store in or over liquid nitrogen.

5. Warming Place a container of sterile water in a water bath at 40°C and leave to equilibrate. Drop ampoules into the warm sterile water.

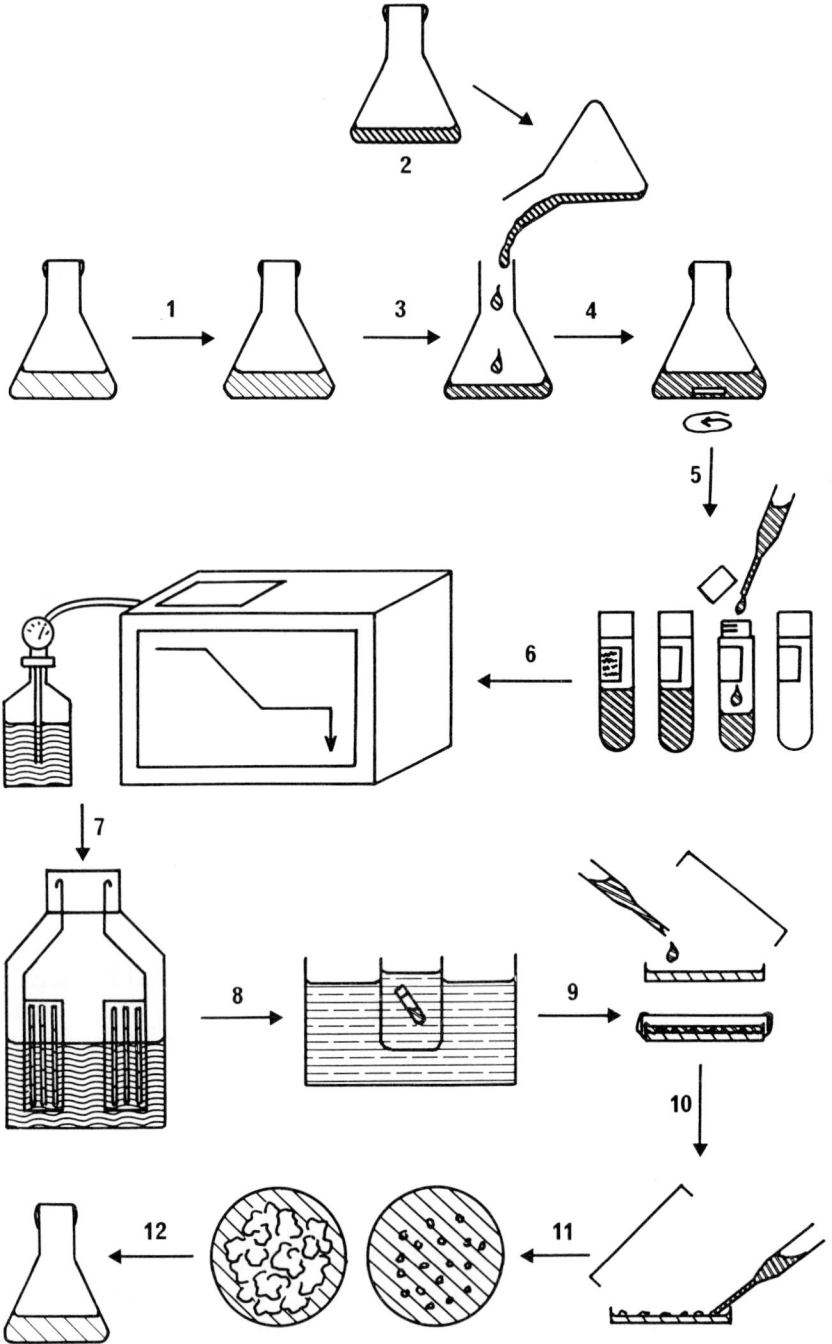

Fig. 1 Cryopreservation procedure for a cell suspension culture; for further details see text, Table II and references (Withers, 1990a; Withers and King, 1980; Withers, 1985c). (1) Transfer cells to pre-growth passage, using modified medium if necessary. (2) Prepare cryoprotectants; cool. (3) Harvest cells; cool. (4) Add cryoprotectants to cells; mix well; incubate. (5) Transfer cryoprotected cell suspension to ampoules; seal and label. (6) Transfer to slow cooling apparatus; cool slowly. (7) Transfer to liquid nitrogen refrigerator. (8) Thaw by warming in water bath. (9) Transfer cells and cryoprotectant solution to semi-solid medium. (10) Remove unincorporated liquid medium. (11) Leave cells to recover growth. (12) Transfer cells to liquid medium. (Reproduced with permission from: Withers,1990a.)

Agitate them until all the ice has just melted, then transfer them to a rack.

6. Post-thaw treatment

Do not wash the cells or dilute the cryoprotectant mixture with fresh medium. Pipette the cell suspension onto a plate of semi-solid medium of normal composition. Approximately 1 ml of suspension will inoculate one or two 5 cm diameter Petri dishes. Incubate under normal conditions.

7. Recovery

After a few days, draw off any liquid that has not been reabsorbed by the cells. After 2–4 weeks the cells should be ready for transfer to liquid medium to resume the normal subculturing cycle.

8. Viability testing

Recovery is the best indicator of viability but in the development of a protocol for an untried specimen it can be helpful to use a viability test such as fluorescein diacetate staining or the TTC test (for details see Withers, 1986).

Refining a sub-optimal procedure

The model procedure described above is widely successful but it would be misleading to suggest that it is the best for all culture systems and species. Table II lists some of the possible variations that might be applied at each stage to increase survival.

Cooling conditions may be adjusted to improve the balance between under- and over-dehydration and minimize ice damage. A procedure that has been found successful involves preparing a quantity of pre-grown, cryoprotected cells and taking samples through a cooling curve as illustrated in Figure 2a. At points on the curve, one ampoule is warmed immediately and another is transferred to liquid nitrogen prior to warming. A viability test is carried out and a comparison of the data (with and without exposure to liquid nitrogen) will reveal the onset of protective dehydration or the onset of any over-dehydration (Fig. 2b).

Once cooling conditions are optimized, attention can be given to other stages in the protocol, for example pre-growth or cryoprotection. It is possible to explore the use of slow thawing; this can in exceptional cases lead to higher post-thaw viability and more rapid recovery (Withers 1980b). With these aspects improved, it is useful to re-examine the cooling curve and further 'tune' the procedure.

Table II: Cryopreservation of a cell suspension culture: alternative treatments that can be applied at different stages of the model procedure should it be unsuccessful.*

PRE-GROWTH

Standard culture medium or medium modified by the addition of one of the following: 0.02 M asparagine; 0.05 M alanine; 0.01 M serine; 0.02 M proline; 5–10% proline; 0.15 M mannitol; 3–6% mannitol; 6–15% sorbitol; 1 M sorbitol; 7.5×10^{-5} M ABA; 5% DMSO; 5–10% trehalose

Duration of exposure to additives: 6 h to 7 days

Temperature: normal or reduced (e.g. 4°C)

CRYOPROTECTANTS

Single: 5–10% DMSO; 5–10% glycerol; 5–10% PEG 6000; 10% proline; 1 M sorbitol; 20% sucrose; 20–40% trehalose

Mixtures: 2.5–10.0% DMSO + 5–10% glycerol; 2.5–10.0% DMSO + 10–20% glycerol + 20% sucrose; 0.5 M DMSO + 0.5 M glycerol + 1 M sucrose; 10–12% DMSO + 5–8% glucose; 10–23% glycerol + 10–17% sucrose; 2 M DMSO + 0.5 M sucrose; 0.5 M DMSO + 0.5 M glycerol + 1 M proline; 1 M DMSO + 0.5 M glycerol + 0.5 M proline; 12% DMSO + 8–10% glucose; 5–10% DMSO + 0.5–1.0 M sorbitol; 10% DMSO + 8% glucose + 10% PEG

Temperature: room temperature or 4°C

Addition: slowly or in one aliquot. Duration of exposure 1 h or, exceptionally, up to 16 h

Transfer to ampoules with 1 ml of cryoprotectant solution. If necessary, allow cells to settle to give a greater cell density

COOLING

Rate: 0.1–2.0°C min^{-1}

Holding temperature, if applied: −23 to −40°C

Holding time: 30 min to 3 h

Transfer temperature: from holding temperature where used, or −23 to −100°C

WARMING

Rapidly in water at 40°C or, exceptionally, slowly (air at room temperature)

POST-THAW TREATMENT

Transfer to recovery medium without washing, or wash one or more times in fresh medium

Table II: *Continued.*

Transfer cells only, or transfer cryoprotectant solution with cells, pipetting away unincorporated solution after 2–7 days

RECOVERY

Medium: semi-solid medium of normal formulation or ammonium-free; normal formulation but with cells supported on a filter paper which is moved to fresh medium after 1 h (for example) then twice daily; semi-solid medium over a feeder cell layer; liquid medium of normal formulation or modified to encourage growth from low cell densities

Growth conditions: normal temperature; normal or reduced lighting

Transfer to fresh liquid medium and resume normal subculturing cycle once vigorous growth has been resumed

* See also Figure 1, Withers (1990) and Withers and King (1980).
Abbreviations: ABA, abscisic acid: DMSO, dimethyl sulphoxide; PEG, polyethylene glycol.

C. Protoplasts

Protoplasts have received less attention than cell suspensions and are of more value as an experimental system to investigate the freezing process than as subjects for conservation. However, the conventional approach to cryopreservation has been used for protoplasts and, more recently, they have been used to explore the potential of vitrification (see Section V.B). This model procedure is based on the conventional approach (see Takeuchi *et al.*, 1982).

Model procedure

1. Pre-growth   Harvest cells in exponential growth and isolate protoplasts.

2. Cryo-
protection   Add 5% (v/v) DMSO + 10% (w/v) glucose; dispense into ampoules.

3. Cooling   Cool at 1 or 2°C per minute to −35°C; transfer to liquid nitrogen.

4. Warming   Warm rapidly in water at 40°C.

5. Post-thaw
treatment   Wash in liquid medium.

6. Recovery   Use standard semi-solid medium for recovery.

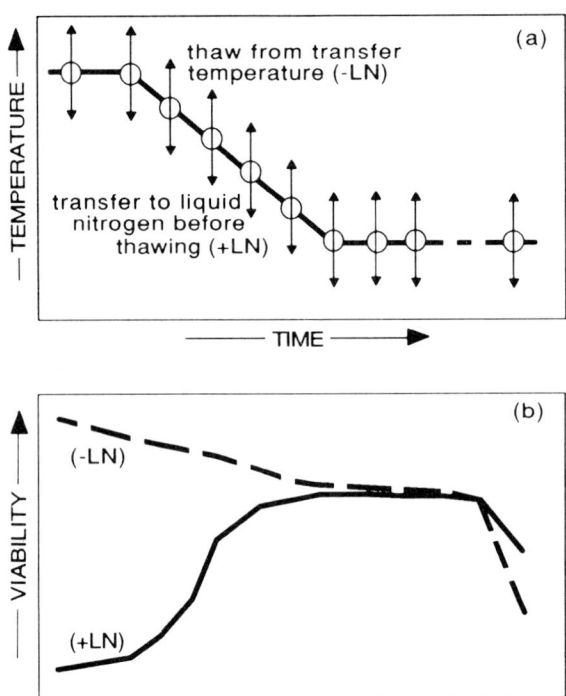

Fig. 2 Strategy for optimizing cryopreservation conditions: (a) Cells are harvested, cryoprotected, transferred to ampoules and then subjected to a cooling regime: for example 1°C per minute to −35°C, then holding for periods of time up to 120 min. At intervals (e.g. before cryoprotection; after cryoprotection; every 5–10 min during slow cooling and holding) ampoules are transferred to liquid nitrogen and then thawed rapidly, corresponding ampoules being thawed directly from the transfer temperature. (b) A viability test (e.g. fluorescein diacetate staining; see Withers, 1986) is carried out on the thawed cells and the two sets of data, with (———) and without (– – –) exposure to liquid nitrogen (LN) are plotted. Ideally, the cells not exposed to liquid nitrogen will show a reduction in viability, slight at first as a result of cryoprotection (second sampling on curve a), and then progressively more severe as cooling proceeds. The cells exposed to liquid nitrogen show poor viability until protective dehydration becomes effective during the slow cooling or holding stages. Over-dehydration will be signified by a drop in viability towards the end of the cooling period. Cells not exposed to liquid nitrogen may at this time show a lower viability, perhaps because of the effects of crystal growth in the early stages of thawing. Inspection of the +LN curve (i.e. cells exposed to liquid nitrogen) indicates the best time to terminate slow cooling by transferring ampoules to liquid nitrogen. If the cells do not follow this idealized behaviour, the shape of the curve can give clues as to the cause. For example, if the cells not exposed to liquid nitrogen lose viability very early during cryoprotection or the first stages of cooling, it is a signal that attention should be given to pre-growth and/or cryoprotection. If viability loss occurs part way through slow cooling, it may be beneficial to reduce the rate or use an intermediate holding temperature. If there is a large gap, throughout the early and middle stages of slow cooling, between the viability of the cells not exposed to liquid nitrogen and those that are, it may indicate that protective dehydration is not occurring and that the holding temperature may need to be lowered or the holding period extended, or that a second, lower holding temperature/period be introduced. This strategy can be extended by the introduction of a comparison between rapid and slow warming, most beneficial for specimens held for relatively long periods of time at the holding temperature and showing signs of overdehydration (see Withers, 1980b).

D. Callus

Callus cultures are very diverse in their morphology. Moreover, they often have characteristics that are known to reduce freeze-tolerance. For example, a callus is often made up of large, dense, slowly growing masses, containing cells of very different sizes and degrees of vacuolation, some being senescent. There may be high levels of autotoxic compounds such as alkaloids. Accordingly, callus cultures can be very difficult to cryopreserve.

Model procedure 1 Most successful procedures have been based upon the principles of cell suspension cryopreservation as described in Section IV.B. With suitable pre-treatment by dissection, gentle homogenization or shaking in liquid medium to reduce the bulk of the callus masses, it is possible to treat the material as a suspension culture (see Section IV.B).

Model procedure 2 This alternative is based on the work of Finkle *et al.* (1982).

1. Pre-growth Harvest actively growing callus.

2. Cryo-protection Add 10% (v/v) DMSO + 10% (w/v) polyethylene glycol (Carbowax 6000) + 8% (w/v) glucose; dispense into ampoules.

3. Cooling Cool at 1°C per minute to −30°C; transfer to liquid nitrogen.

4. Warming Warm rapidly in water at 40°C.

5. Post-thaw treatment Wash in liquid medium at 22°C.

6. Recovery Use standard semi-solid medium for recovery.

E. Shoot cultures

Shoot cultures *per se* cannot be cryopreserved; they are too bulky, complex and heterogeneous. However, shoot tips or meristem tips dissected from shoots can be cryopreserved.

There are two very different approaches to cryopreservation, one using very rapid cooling and the other slow, dehydrative cooling; both procedures are described below. In general, slow cooling is more successful, although it is clear from research into cooling rates, with or without seeding to initiate ice formation at different temperatures, that shoot-tips and meristems are extremely vulnerable to cryoinjury.

The development of cryopreservation procedures by several different groups of scientists means that there are many variations to consider, particularly in the case of the slow cooling

Model Procedure 2 (see below). These are listed in Table III, which excludes variations at the recovery stage where knowledge of species and culture requirements is critical. The culture medium may be modified to stimulate a particular growth pattern, often with growth regulator combinations that promote cell recovery in the damaged shoot tip, at the same time risking disorganized growth (see also Section V.A).

Model procedure 1   This is based on the work of Grout and Henshaw (1978) and the subsequent experience of the present author and colleagues (e.g. Benson *et al.*, 1989; Withers *et al.*, 1988).

1. Pre-growth	Dissect shoot/meristem tips from a healthy, relatively rapidly growing parent plant/culture; pre-grow for 2 days supported on a filter paper in contact with standard liquid medium or medium supplemented with 5% (v/v) DMSO.
2. Cryo-protection	Add 5% (v/v) DMSO; dispense into ampoules.
3. Cooling	Collect each specimen separately on the end of a sterile hypodermic needle and plunge into liquid nitrogen, or place a small number of specimens in a drop of cryoprotectant in an ampoule and drop it into liquid nitrogen.
4. Warming	For specimens on a needle, swirl in liquid medium at room temperature; for ampoules, use water at 40°C.
5. Post-thaw treatment	No treatment is required unless cryoprotectant toxicity is suspected, in which case wash specimens in standard liquid medium.
6. Recovery	Use standard semi-solid medium for recovery.

Model procedure 2   This is a simple method based on the work of Kartha *et al.* (1979) and should be used as a framework for exploring the variations offered in Table III.

1. Pre-growth	Dissect shoot/meristem tips from a healthy rapidly growing parent plant/culture; pre-grow for 2 days supported on a filter paper in contact with standard liquid medium or medium supplemented with 5% (v/v) DMSO.
2. Cryo-protection	Add 5% (v/v) DMSO; dispense into ampoules.

3. Cooling Cool at 0.5°C per minute to −40°C; transfer to liquid nitrogen.

4. Warming Rapid in water at 40°C.

5. Post-thaw No treatment is required unless cryoprotectant toxicity is sus-
treatment pected, in which case wash specimens in standard liquid medium.

6. Recovery Use standard semi-solid medium for recovery.

Table III: Cryopreservation of shoot/meristem tip cultures: alternative
treatments that can be applied at different stages of the slow-cooling based
procedure.

PRE-GROWTH

Use an independently growing plant or a culture as source of explants

Plant/culture growth conditions: normal or reduced temperature to cold
harden (e.g. 0 to −1°C for 1–8 weeks; reduced temperature 24 h per day or
alternating normal daytime temperature and reduced night temperature)

Explant: vary size and/or number of leaf primordia; pre-grow explant on
normal medium or medium modified by the addition of: 4–5% DMSO; 1 M
sorbitol; 0.75 M sucrose

Duration of pre-growth: 0–4 days

Temperature: normal or reduced (e.g. 4°C)

CRYOPROTECTANTS

Single: 4–16% DMSO; 5–10% glycerol

Mixtures: 5% DMSO + 5% glycerol; 5% DMSO + 5% glycerol + 5% sucrose;
10% glycerol + 10% glucose + 10% PEG; 10% DMSO + 5% glucose; 10%
DMSO + 5% sucrose; 10% DMSO + 2 M sorbitol. Exceptionally, shoot tips
can survive without cryoprotection

Temperature: room temperature or 4°C

Addition: slowly or in one aliquot

Duration of exposure: 1 h or, exceptionally, up to 16 h

Transfer to ampoules without cryoprotectant solution or with 1 drop to 1 ml
of solution; place directly in ampoule or supported on a piece of tissue;
alternatively, place in drops of cryoprotectant solution on a sheet of
aluminium foil in a Petri dish

Table III: *Continued.*

COOLING

Rate: 0.2–1.0°C min^{-1} with or without seeding by, for example, touching liquid nitrogen-cooled forceps against ampoule

Transfer temperature: −30 to −40°C

WARMING

Ampoules: rapidly in water at 40°C or, exceptionally, slowly (air at room temperature)

Specimens on tissue or foil carrier: drop into liquid culture medium at 20°C

POST-THAW TREATMENT

Transfer to semi-solid recovery medium without washing or wash one or more times in fresh medium

RECOVERY

Medium: semi-solid medium of normal formulation or supplemented with growth regulators according to species

Growth conditions: normal temperature; normal or reduced lighting

Transfer to fresh medium and resume normal subculturing cycle once vigorous growth has been resumed

Abbreviations: DMSO, dimethyl sulphoxide; PEG, polyethylene glycol.

F. Somatic embryos

The process of somatic embryogenesis is becoming increasingly important in all aspects of tissue culture from mass propagation to genetic manipulation, secondary product synthesis and genetic conservation. It is an exciting culture system. There is interest in storing embryogenic cultures for their intrinsic value and also as alternatives to the more difficult manipulation of shoot cultures (see Section V.A).

In the context of plant genetic conservation, somatic embryos have an additional important application. There is a need to develop satisfactory storage methods for recalcitrant seeds. Cryo-preservation is an option, but many recalcitrant seeds are very large, dense, highly hydrated and sensitive to dehydration and/or low temperatures. The zygotic embryo within the seed may also be too large and otherwise unsuitable for cryopreservation, but it may be stimulated to produce somatic embryos that are signifi-cantly smaller.

Research has been carried out on the development of cryo-preservation procedures for somatic embryos and embryogenic cultures with great success. Because of the diversity of approaches linked to the stage of development of the material in question, ranging from a cell suspension containing embryogenic cell masses, through more bulky, embryogenic callus-containing early-stage embryo structures, to fully formed advance-stage embryos and plantlets, three different model procedures are described.

Model procedure 1 — for embryogenic cell suspensions The procedure offered in Section IV.B could be tried. Alternatively, the following is a further example developed for an embryogenic suspension (see Kartha *et al.*, 1988).

1. Pre-growth	Culture for 24 h in liquid medium containing 0.4 M sorbitol.
2. Cryo-protection	Add 5% (v/v) DMSO + 0.4 M sorbitol; dispense into ampoules.
3. Cooling	Cool at 0.3°C per minute to −35°C; transfer to liquid nitrogen.
4. Warming	Rapid in water at 37°C.
5. Post-thaw treatment	Place unwashed cells on a filter paper over standard semi-solid medium; transfer the filter paper to fresh medium after 1 h and 18 h.
6. Recovery	Use standard semi-solid medium, but at a low light intensity, for recovery.

Model procedure 2 — for embryogenic callus and early stage embryos This is based on the work of Engelmann *et al.* (Engelmann, 1991; Engelmann *et al.*, 1985). It differs from other procedures in that there is no cryoprotection stage and the cooling rate is not critical.

1. Pre-growth	Grow cultures on medium containing 0.3 M sucrose for 2 months. Transfer embryogenic masses to medium containing 0.75 M sucrose for 7 days.
2. Cryo-protection	No addition of cryoprotectants; dispense embryogenic masses into ampoules.
3. Cooling	Cool rapidly by direct immersion in liquid nitrogen, or slowly at the rate of 0.5°C to 40°C per minute to −100°C; transfer to liquid nitrogen.
4. Warming	Warm rapidly in water at 40°C.
5. Post-thaw treatment	No post-thaw treatment is needed.

6. Recovery	Culture for 7 days on medium containing 0.3 M sucrose, followed by 2 weeks on medium containing 0.1 M sucrose, before transfer to standard medium.

Model procedure 3 — for fully formed embryos This procedure, termed 'dry freezing', is taken from Withers (1979), and is one of the exceptional cases where slow thawing is beneficial. It avoids physical injury to the very brittle frozen embryos/plantlets.

1. Pre-growth	Harvest embryos from standard culture conditions.
2. Cryo-protection	Add 10% (v/v) DMSO; blot dry of superficial moisture using sterile filter paper; transfer to sterile aluminium foil envelopes.
3. Cooling	Cool at 1°C per minute to −40°C; transfer to liquid nitrogen.
4. Warming	Warm slowly in air.
5. Post-thaw treatment	No post-thaw treatment is needed.
6. Recovery	Use standard semi-solid medium for recovery.

G. Cryopreservation equipment

The scale of the culture storage operation and its objectives will be considerations in the choice of equipment. If the work in prospect is to carry out large-scale storage in routine application where exact reproducibility is essential, to develop cryopreservation procedures for previously untried material, or to carry out fundamental investigations into the cryopreservation process, then it is clear that a high level of sophistication will be required. In this case, there is no substitute for a programmable apparatus, cooled by liquid nitrogen, available from several manufacturers. Where a more modest operation is in prospect and/or where exact reproducibility is less essential, then it is possible to improvise by using an electrically cooled alcohol bath, or an insulated container at −70°C. For further details, see Withers (1984) and Withers and King (1980).

All the above points apply to slow cooling. For rapid cooling or for cooling rate-independent work, the demands are obviously less. In all cases it is necessary to have a purpose-built liquid nitrogen refrigerator, preferably more than one to allow for back-up duplication. For a large-scale operation, a liquid nitrogen level sensor and topping-up facilities are an advantage. A warm water bath is required for most thawing procedures.

V. CONTINUING PROBLEMS AND RESEARCH CHALLENGES IN CRYOPRESERVATION

There are few serious problems in the cryopreservation of cell suspension cultures and, although callus cultures are more difficult to handle, the major problems now lie with organized cultures, particularly shoot tips. The achievement of cryopreservation conditions to suit the diverse cell types and to maintain the continuity of the symplasm, and the application of recovery conditions that promote regeneration of a normal structure without incurring the risk of somaclonal variation from excessive disorganized growth or callusing represent the greatest current challenge to the plant cryobiologist.

Ways of optimizing existing knowledge have been discussed and can be supplemented by the use of such techniques as cryomicroscopy (McClellan et al., 1990). An alternative approach is 'vitrification'. This procedure was originally developed for animal embryos where survival of the entire specimen was a critical objective and has now been extended to plant material. It promises to extend the scope of in vitro conservation significantly. It is unlikely to supersede slow cooling/dehydration based cryopreservation, rather it is seen as an important alternative with special applications. Its most notable characteristics are the use of complex cryoprotectant mixtures at very high, potentially toxic concentrations, and flexibility in cooling rates. Great care is needed at the stages of loading and unloading cryoprotectants to avoid damage. A question unanswered as yet is the level of risk of devitrification during storage. Successes so far with plant material include cell suspension cultures, somatic embryos and protoplasts (Langis and Steponkus, 1990; Langis et al., 1989; Uragami et al., 1989). At present it would be misleading to recommend a model procedure, but the reader is encouraged to follow developments closely.

There is also scope for new, imaginative variations of existing approaches, drawing on knowledge from other disciplines and technical developments within plant tissue culture. An example in the former category would be the wider use of hardening treatments based on natural phenomena, such as the application of sugars or amino acids produced in nature under conditions of stress (e.g. Engelmann, 1990b; Withers and King, 1980), or the application of low temperatures during pregrowth (e.g. Reed, 1989). Greater attention could also be paid to the culture environment, and recent studies by Benson et al. (1989) would testify to the importance of, for example, lighting conditions.

Developments in general plant tissue culture would suggest the application of artificial seed technology (Redenbaugh, 1990), as a means of either handling material during cryopreservation or understanding and avoiding dehydration stresses.

A. Stability in cryopreserved cultures

It has already been suggested that slow growth is sub-optimal in the stability that it offers. For unorganized cultures this mainly applies to risks of somaclonal variation and genetic instability. For organized cultures, the risks appear to be to viability, vigour and freedom from contamination. Can cryopreservation offer better? On the basis of current evidence cryopreservation is a safe procedure (see Withers, 1986; 1987a) with regard to the production of specific secondary compounds, resistance to anti-metabolites or the expression of totipotency. Some suggestion of increased tolerance to cryopreservation upon repeated exposure (Watanabe *et al.*, 1985) may indicate instability, but is more likely to indicate selection, possibly on the basis of morphology rather than genotype. In any case, frequently repeated cryo-preservation is unlikely to be necessary in practice.

Intentional 'cryoselection' under conditions designed to give minimal survival is another matter. Its prospects for producing cold-tolerant genotypes look promising (K. K. Kartha, personal communication, 1990).

VI. *IN VITRO* CONSERVATION IN PRACTICE

There are now a number of examples of the routine application of slow growth procedures to shoot cultures and cryopreservation to cell suspension cultures to give credibility to *in vitro* storage (e.g. see Di Maio and Shillito, 1989; Withers, 1991a,b). Important as storage is, however, it is only part of a conservation system that begins with the acquisition of germplasm by either collecting in the field or transferring from another collection, and leads eventually to germplasm exchange. There is a role for the *in vitro* approach at each stage in the system: collecting material in the field using simple, *in vitro* collecting procedures, virus eradica-tion to clean up germ plasm, disease indexing, multiplication, storage *per se*, and *in vitro* exchange to distribute germ plasm. For further details on the development and application of complete conservation systems, the reader is referred to recent reviews (Withers, 1989; 1991a,b).

In vitro gene banks require safe and efficient conservation, of material that should be characterized, documented, adequately replicated and duplicated, and conserved under high standards of management. In recognition of the importance of this, the International Board for Plant Genetic Resources (IBPGR) and CIAT undertook an experiment commencing in early 1987 and now nearing completion to test standards of replication and other variables (Chavez *et al.*, 1987; IBPGR, 1986). This pilot '*In vitro* active genebank' experiment involved the introduction of replicates of 100 test genotypes (from the 5000 held at CIAT in a field gene bank) into culture via disease indexing and thermotherapy. Assessment of the workload involved in the pilot exercise made it clear that a rationalization would be necessary to meet the conflicting demands of material resources and germ plasm security.

As yet, little attention has been given to the development of management procedures for cryopreserved collections of plant cultures. Some guidelines are suggested by Di Maio and Shillito (1989), and experience with other biological organisms and a knowledge of plant tissue culture in general will guide the user.

REFERENCES

Advisory Committee on Dangerous Pathogens (1990). *Categorisation of Pathogens According to Hazard and Categories of Containment.* Second edition, Her Majesty's Stationery Office, London.

Agereau, J. M., Courtois, D. and Petiard, V. (1986). *Plant Cell Rep.* 5, 372–376.

Alexander, M., Daggett, P.-M., Gherna, R., Jong, S., Simione, F. and Hatt, H. (1980). *American Type Culture Collection Methods I. Laboratory Manual on Preservation Freezing and Freeze-drying.* Rockville, Maryland, American Type Culture Collection.

Annear, D. I. (1956). *J. Hyg., Camb.* 54, 487–508.

Annear, D. I. (1958). *Aust. J. Exp. Biol. med. Sci.* 36, 211–221.

Annear, D. I. (1962). *J. Gen. Microbiol.* 27, 341–343.

Anon. (1972–1981). *Index of Fungi.* CAB International Mycological Institute (CMI), London.

Anon. (1982). Mites. CAB International Mycological Institute (CMI), London.

Antheunise, J. (1972). *Anton. van Leeuwenhoek* 38, 617–622.

Arora, R. and Bhojwani, S. S. (1989). *Plant Cell Rep.* 8, 44–47.

Ashwood-Smith, M. J. and Farrant, J. (eds) (1980). *Low Temperature Preservation in Medicine and Biology.* Pitman Medical, Tunbridge Wells.

Atkinson, R. G. (1954). *Can. J. Bot.* 32, 673–678.

Balch, W. E. and Wolfe, R. S. (1976). *Appl. Environ. Microbiol.* 32, 781–791.

Balch, W. E., Fox, G. E., Magrum, L. J., Woese, C. R. and Wolfe, R. S. (1979). *Microbiol. Rev.* 43, 260–290.

Banerjee, N. and DeLanghe, E. A. L. (1985). *Plant Cell Rep.* 4, 351–354.

Banno, T. and Sakane, T. (1979). *IFO Res. Comm.* 9, 35–45.

Barnes, E. M. (1969). In *Methods in Microbiology* (J. R. Norris and D. W. Ribbons, eds), Vol. 3B, pp. 151–160. Academic Press, New York.

Barnes, E. M. and Impey, C. S. (1971). In *Isolation of Anaerobes* (D. A. Shapton and R. G. Board, eds), pp. 115–132, SAB Technical Series No. 5. Academic Press, London.

Barnes, E. M. and Impey, C. S. (1974). *J. Appl. Bact.* 37, 393–409.

Barnes, E. M. and Impey, C. S. (1978). In *Techniques for the Study of Mixed Populations* (D. W. Lovelock and R. Davies, eds), pp. 89–105, SAB Technical Series No. 11. Academic Press, London.

Barrett, R. W., Johnson, G. B. and Ogata, W. N. (1965). *Genetics, Princeton* 52, 233–246.

Baserga, R. (1989). In *Cell Growth and Division* (R. Baserga, ed.), IRL Press, Oxford.

Bassel, J., Contopoulou, R., Mortimer, R. and Fogel, S. (1977). *UK Federation for Culture Collections Newsletter* No. 4, p. 7.

Baynes, S. M., Emerson, L. and Scott, A. P. (1979). *Fish. Res. Tech. Rep., MAFF Direct. Fish. Res. Tech. Res., Lowestoft*, **53**, 13–18.

Belay, N., Sparling, R. and Daniels, L. (1984). *Nature (London)* **312**, 286–288.

Ben-Amotz, A. and Gilboa, A. (1980). *Marine Ecol. Progr. Ser.* **2**, 157–161.

Benson, E. E. (1990). *Free Radical Damage in Stored Plant Germplasm*. IBPGR, Rome.

Benson, E. E., Harding, K. and Smith, H. (1989). *Cryoletters* **10**, 323–344.

Berger, L. R. (1970). In *Proceedings of the First International Conference on Culture Collections* (H. Iizuka and T. Hasegawa, eds), pp. 265–267. University of Tokyo Press, Tokyo.

Berny, J. F. and Hennebert, G. L. (1989). *Abstracts, Meeting of Contractors, Biotechnology Action Programme of the Commission of the European Community's Biotechnology Division*.

Biebl, H. and Malik, K. A. (1976). In *Proceedings of the Second International Symposium on Photosynthetic Prokaryotes* (G. A. Codd and W. D. P. Stewart, eds), pp. 31–33. Dundee.

Biebl, H. and Pfennig, N. (1981). In *The Prokaryotes* (M. P. Starr, H. Stolp, H. G. Trüper, A. Balow and H. G. Schlegel, eds), pp. 267–273. Springer Verlag, Berlin.

Boeswinkel, H. J. (1976). *Trans. Br. Mycol. Soc.* **66**, 183–185.

Bollinger, R. O., Mussalam, N. and Stuhlberg, C. S. (1974). *J. Parasitol.* **60**, 368–369.

Booth, C. (1971a). *Methods Microbiol.* **4**, 49–94.

Booth, C. (1971b). *The Genus* Fusarium. Commonwealth Mycological Institute, Kew.

Bose, S. K. (1963). In *Bacterial Photosynthesis* (H. Guest, A. san Pietro and L. P. Vernon, eds), pp. 501–519. Antioch Press, Yellow Springs, Ohio.

Bridgen, M. P. and Staby, G. L. (1983). In *Handbook of Plant Cell Culture, Volume 1, Techniques for Propagation and Breeding* (D. A. Evans, W. R. Sharp, P. V. Ammirato and Y. Yamada, eds), pp. 816–827. Macmillan, New York.

Brock, T. D. and O'Dea, K. (1977). *Appl. Environ. Microbiol.* **33**, 254–256.

Brown, M. R., Jeffery, S. W. and Garland, C. D. (1989). *CSIRO Mar. Lab. Rep.* **205**, (vi), 44.

Brown, M. R., Jeffrey, S. W., Nichols, P. D., Rogers, G. I. and Garland, C. D. (1989). *J. Exp. Mar. Biol, Ecol.* **128**, 219–240.

Bryant, M. P. and Robinson, I. M. (1961). *J. Dairy Sci.* **44**, 1446–1456.

Buell, C. B. and Weston, W. H. (1947). *Am. J. Bot.* **34**, 555–561.

Butterfield, W., Jong, S. C. and Alexander, M. J. (1974). *Can. J. Microbiol.* **20**, 1665–1673.

Butterfield, W., Jong, S. C. and Alexander, M. J. (1978). *Mycologia* **70**, 1122–1124.

Calcott, P. H. (1978). *Freezing and Thawing Microbes.* Meadowfield Press, UK.

Calcott, P. H. (1982). *Ohio J. Sci.* **82**, 56–66.

Calcott, P. H. and Gargett, A. M. (1981). *Fems Microbiol. Lett.* **10**, 151–155.

Calcott, P. H., Wood, D. and Anderson, L. (1981). *Cryoletters* **4**, 99–106.

Caldwell, D. R. and Bryant, M. P. (1966). *Appl. Microbiol.* **14**, 794–801.

Callow, L. L. and Farrant, J. (1973). *Int. J. Parasitol.* **3**, 77–88.

Caplin, S. M. (1959). *Am. J. Bot.* **46**, 324–329.

Carmichael, J. W. (1956). *Mycologia* **48**, 378–381.

Carmichael, J. W. (1962). *Mycologia* **54**, 432–436.

Castellani, A. (1939). *J. Trop. Med. Hyg.* **42**, 225–226.

Castellani, A. (1967). *J. Trop. Med. Hyg.* **70**, 181–184.

Castenholz, R. W. (1988). In *Methods in Enzymology* (L. Packer and A. N. Glazer, eds), pp. 68–93. Academic Press, New York.

Castenholz, R. W. and Pierson, B. K. (1981). In *The Prokaryotes* (M. P. Starr, H. Stolp, H. G. Trüper, A. Balow and H. G. Schlegel, eds), pp. 290–298. Springer Verlag, Berlin.

Challen, M. P. and Elliot, T. J. (1986). *J. Microbiol. Methods* **5**, 11–23.

Chavez, R., Roca, W. M. and Williams, J. T. (1987). *FAO/IBPGR Plant Genetic Resources Newsletter* **71**, 11–13.

Clark, G. and Dick, M. W. (1974). *Trans. Br. Mycol. Soc.* **63**, 611–612.

Coe, A. W. and Clark, S. P. (1966). *Mon. Bull. Minist. Hlth.* **25**, 97–100.

Coggeshall, L. T. (1939). *Proc. Soc. Exp. Biol. Med.* **42**, 499–501.

Coghlan, J. D., Lumsden, W. H. R. and McNeillage, G. J. C. (1967). *J. Hyg., Camb.* **65**, 373–79.

Cowan, S. T. (1974). *Cowan and Steel's Manual for the Identification of Medical Bacteria*, 2nd edn. Cambridge University Press, Cambridge.

Cowling, A. J. (1986). *Protistologica* **22**, 181–191.

Cowling, A. J. (1991). In *The Biology of Free Living Heterotrophic Flagellates* (D. J. Patterson and J. Larsen, eds), Systematics Association, special volume, Oxford University Press, Oxford, UK. In press.

Crane, J. and Hughes, H. (1990). *Hort. Science* **25**, 794–795.

Crush, J. R. and Pattison, A. C. (1975). In *Endomycorrhizas* (F. E.

Sanders, B. Mosse and P. B. Tinker (eds), pp. 485–509. Academic Press, London.

Cunningham, M. P., Lumsden, W. H. R. and Webber, W. A. F. (1963). *Exp. Parasitol.* **14**, 280–284.

Cunningham, M. P., Van Hoeve, K. and Grainge, E. B. (1964). *East African Trypanosomiasis Research Organization Annual Report, July 1963–December 1964*, pp. 26–30.

Cunningham, M. P., Brown, C. G. D., Burridge, M. J. and Purnell, R. E. (1973). *Int. J. Parasitol.* **3**, 583–587.

Dade, H. A. (1960). In *Herb Handbook*, pp. 78–83. Commonwealth Mycological Institute, Kew.

Dagliesh, R. J. and Mellors, L. T. (1974). *Int. J. Parasitol.* **4**, 169–172.

Dagliesh, R. J., Mellors, L. T. and Blight, G. W. (1980). *Cryobiology* **17**, 410–417.

Day, J. G. (1987). Ph.D. Thesis, Dundee University, UK.

Day, J. G., Priestly, I. M. and Codd, G. A. (1987). In *Plant and Animal Cells, Process Possibilities* (C. Webb and F. Mavituna, eds), pp. 257–261. Ellis Horwood, Chichester.

Day, J. G., Edward, A. P. and Rodgers, G. A. (1991). In *Biomass: Recent advances in Algal Biotechnology* (A. Richmond and A. Vorshck, eds), Elsevier Science Publishers, Barking, UK. In press.

De Pauw, N. and Pruder, G. (1986). In *Realism in Aquaculture: Achievements, Constraints, Perspectives*, pp. 77–106. European Aquaculture Society, Bredane, Belgium.

Deus-Neumann, B. and Zenk, M. H. (1984). *Planta Medica* **50**, 427–431.

Diamond, L. S., Harlow, D. R. and Cunnick, C. C. (1978) *Trans. R. Soc. Trop. Med. Hyg.* **72**, 431–432.

Dietz, A. (1975). In *Round Table Conference on Cryogenic Preservation of Cell Cultures* (A. P. Rinfret and A. B. La Salle, eds), pp. 22–36. National Academy of Science, Washington, DC.

Diffley, P., Honigberg, B. M. and Mohn, F. A. (1976). *J. Parasitol.* **62**, 136–137.

Diller, K. R. and Knox, J. M. (1983). *Cryoletters*, **5**, 239–254.

Di Maio, J. J. and Shillito, R. D. (1989). *J. Tissue Cult. Methods* **12**, 163–169.

Doory, Y. Al- (1968). *Mycologia* **60**, 720–723.

Doyle, A., Morris, C. B. and Armitage, W. J. (1989a). In *Advances in Biotechnological Processes*, Volume 7 (A. Mizrahi, ed.). Alan R. Liss Inc., New York.

Doyle, A., Morris, C. B. and Melling J. M. (1989b). In *Biotechnology and Genetic Engineering Reviews*, Volume 7 (G. E. Russell and M. P. Tombs, eds), pp. 259–280. Intercept Ltd, Andover.

Doyle, A., Kirsop, B. E. and Hay, R. (eds) (1990). *Living Resources for Biotechnology: Animal Cells*. Cambridge University Press, Cambridge.

Droop, M. R. (1961). *Rev. Algolog.* **5**, 249–259.

Droop, M. R. (1967). *Br. Phycol. Bull.* **15**, 295–297.

Druart, P. H. (1985). In *In Vitro Techniques — Propagation and Long-Term Storage* (A. Schafer-Menuhr, ed.), pp. 167–171. Nijhoff/Junk for CEC, Dordrecht.

Dwyer, D. M. and Honigberg, B. M. (1971). *J. Parasitol.* **57**, 190–191.

Eagle, H. (1955). *J. Biol. Chem.* **214**, 839.

Earle, W. R., Shilling, E. L., Stark, T. H., Straus, N. P., Brown, M. F. and Shelton, E. (1943). *J. Natl Cancer Inst.* **4**, 165–212.

Ellinghausen, H. C. and McCullough, W. G. (1965a). *Am. J. Vet. Res.* **26**, 39–44.

Ellinghausen, H. C. and McCullough, W. G. (1965b). *Am. J. Vet. Res.* **26**, 45–51.

Elliott, J. J. (1976). *Trans. Br. Mycol. Soc.* **67**, 545–546.

Ellis, J. J. (1979). *Mycologia* **71**, 1072–1075.

Engel, J. C., Perez, A. C. and Wynne de Martini, G. (1980). *Medicina (Buenos Aires)* **40**, 103–108.

Engelmann, F. (1990a). *Compt. Rend. Acad. Sci.* **311** (Série III), 679–684.

Engelmann, F. (1990b). *Bull. Soc. Bot. Francais*, in press.

Engelmann, F., Duval, Y. and Derueddre, J. (1985). *Compt. Rend. Acad. Sci.* **301** (Série III), 111–116.

Eyles, D. E., Coleman, N. and Cavanaugh, G. J. (1956). *J. Parasitol.* **42**, 408–413.

Farri, T. A., Warhurst, D. C. and Marshall, T. F. (1983). *Trans. R. Soc. Trop. Med. Hyg.* **77**, 259–266.

Feltham, R. K. A., Power, A. K., Pell, P. A. and Sneath, P. H. A. (1978). *J. Appl. Bacteriol.* **44**, 313–316.

Fennell, D. I. (1960). *Bot. Rev.* **26**, 79–141.

Figueiredo, M. B. (1967). *O Biologico* **33**, 9–15.

Figueiredo, M. B. and Pimentel, C. P. V. (1975). *Summa Phytopathol.* **1**, 299–302.

Finkle, B. J., Ulrich, J. M. and Tisserat, B. (1982). In *Plant Cold Hardiness and Freezing Stress: Volume 2: Mechanisms and Crop Implications* (P. H. Li and A. Sakai, eds), pp. 643–660. Academic Press, New York.

Finlay, B. F., Rogerson, M. and Cowling, A. J. (1988). *A Beginner's Guide to the Collection, Isolation, Cultivation and Identification of Freshwater Protozoa.* Culture Collection of Algae and Protozoa.

Franks, F., Mathias, S. F., Galfre, P., Webster, S. D. and Brown, D. (1983). *Cryobiology* **20**, 298–309.

Freshney, R. I. (1987). *Culture of Animal Cells: a Manual of Basic Techniques*, 2nd ed. Alan R. Liss Inc, New York.

Fry, R. M. (1954). In *Biological Applications of Freezing and Drying* (R. J. C. Harris, ed.), pp. 215–252. Academic Press, New York.

Fry, R. M. (1966). In *Cryobiology* (H. T. Meryman, ed.), pp. 665–696. Academic Press, London.

Fry, R. M. and Greaves, R. I. N. (1951). *J. Hyg., Camb.* **49**, 220–246.

Gale, A. W., Schmitt, C. G. and Bromfield, K. R. (1975). *Phytopathology* **65**, 828–829.

Giese, A. C. (1973). *Blepharisma: The Biology of a Light-sensitive protozoan.* Stanford University Press, Stanford.

Gilmour, M. N., Turner, G., Berman, R. G. and Kreuzer, A. K. (1978). *Appl. Environ. Microbiol.* **35**, 84–88.

Gold, K., Pfister, R. M. and Liguori, R. (1970). *J. Protozool.* **17**, 210–212.

Goldie-Smith, E. K. (1956). *J. Elisha Mitchell Sci. Soc.* **72**, 158–166.

Gordon, W. L. (1952). *Can. J. Bot.* **30**, 209–251.

Griffiths, J. B. (1985). In *Animal Cell Biotechnology*, Volume 2 (R. E. Spier and J. B. G. Griffiths (eds). pp. 3–12. Academic Press, London.

Griffiths, J. B. (1986). In *Animal Cell Culture, a Practical Approach* (R. I. Freshney, ed.). pp. 1–17. IRL Press, Oxford.

Grivell, A. R. and Jackson, J. F. (1969). *J. Gen. Microbiol.* **58**, 423–425.

Grout, B. W. W. (1990). In *Progress in Plant Cellular and Molecular Biology* (H. J. J. Nijkamp. L. H. W. van der Plas and J. van Aartrijk, eds), pp. 13–22. Kluwer, Dordrecht.

Grout, B. W. W. and Henshaw, G. G. (1978). *Ann. Bot.* **42**, 1227–1229.

Grout, B. W. W. and Morris, J. G. (eds) (1987). *The Effects of Low Temperature on Biological Systems.* Edward Arnold, London.

Hartsell, S. E. (1956). *Appl. Microbiol.* **4**, 350–355.

Hatt, H. D. (ed.) (1980). *American Type Culture Collection Methods. 1. Laboratory Manual on Preservation, Freezing and Freeze-Drying.* American Type Culture Collection, Rockville, MD.

Hawksworth, D. L., Sutton, B. C. and Ainsworth, G. (1983). *Ainworth's and Bisby's Dictionary of Fungi*, 8th edn. Commonwealth Mycological Institute, Kew.

Heaf, D. D. and Lee, D. (1971). *J. Gen. Microbiol.* **68**, 249–251.

Heckly, R. J. (1961). *Adv. Appl. Microbiol.* **3**, 1–76.

Heckly, R. J. (1978). *Adv. Appl. Microbiol.* **24**, 1–53.

Helm, M. M., Laing, I. and Jones, E. (1979). *Fish. Res. Tech. Rep., MAFF Direct. Fish, Res., Lowestoft*, **53**, 1–7.

Henshaw, G. G. (1975). In *Crop Genetic Resources for Today and Tomorrow* (O. H. Frankel and J. G. Hawkes, eds), pp. 349–358. Cambridge University Press, Cambridge.

Hieda, K. (1981). *Mutation Res.* **84**, 17–27.

Hill, L. R. (1981). In *Essays in Applied Microbiology* (J. R. Norris and M. H. Richmond, eds). pp. 2/1–2/31. John Wiley & Sons, Chichester.

Hill, L. R., Kocur, M. and Malik, K. A. (1990). In *Living Resources for Biotechnology: Bacteria* (L. R. Hill and B. E. Kirsop, eds), pp. 62–80. Cambridge University Press, Cambridge.

Hilpert, R., Winter, J., Hammes, W. and Kandler, O. (1981). *Zbl. Bact. Hyg., I. Abt. Orig. C* **3**, 149–160.

Hippe, H. and Tilly, I. (1982). *XIIIth Int. Congr. Microbiol.*, Boston, Mass., USA. Abstr. P 54.

Hiraoka, N. and Kodama, T. (1984). *Plant Cell, Tissue and Organ Culture* **3**, 349–357.

Holdeman, L. V., Cato, E. P. and Moore, W. E. C. (1977). *Anaerobe Laboratory Manual*, 4th edn. Anaerobe Laboratory, Virginia Polytechnic Institute and State University, Blacksburg, Virginia.

Hollingdale, M. R., Leland, P., Sigler, C. I. and Leef, J. L. (1985). *Trans. R. Soc. Trop. Med. Hyg.* **79**, 206–208.

Holm-Hansen, O. (1964). *Can. J. Bot.* **42**, 127–137.

Holm-Hansen, O. (1967). *Cryobiology* **4**, 17–23.

Holm-Hansen, O. (1973). In *Handbook of Phycological Methods* (J. R. Stein, ed.), pp. 195–206. Cambridge University Press, Cambridge.

Hubalek, Z. and Kochova-Kratochvilova, A. (1978). *Antonie van Leeuwenhoek* **44**, 229–241.

Hungate, R. E. (1950). *Bact. Rev.* **14**, 1–49.

Hungate, R. E. (1969). In *Methods in Microbiology* (J. R. Norris and D. W. Ribbons, eds), Volume IIIB, pp. 117–132. Academic Press, New York.

Hungate, R. E. (1966). *The Rumen and its Microbes*. Academic Press, London.

Hunt, G. A., Gourevitch, A. and Lein, J. (1958). *J. Bact.* **76**, 453–454.

Hwang, S.-W. (1960). *Mycologia* **52**, 527–529.

Hwang, S.-W. (1966). *App. Microbiol.* **14**, 784–788.

Hwang, S.-W. (1968). *Mycologia* **60**, 613–621.

Hwang, S.-W., Kwolek, W. F. and Haynes, W. C. (1976). *Mycologia* **68**, 377–387.

IBPGR (1986). *Design, Planning and Operation of* in vitro *Genebanks*. IBPGR, Rome.

IBPGR-CIAT (1991). *IBPGR-CIAT Pilot* in vitro *Active Genebank*. In press.

Imhoff, J. F., Trüper, H. G. and Pfennig, N. (1984). *Int. J. Syst. Bacteriol.* **34**, 340–343.

Ivey, M. H. (1975). *J. Parasitol.* **61**, 1101–1103.

Jackson, N. E., Miller, R. H. and Franklin, R. E. (1973). *Soil Biol. Biochem.* **5**, 205–212.

Jaworski, G. H. M., Wiseman, S. W. and Reynolds, C. S. (1988). *Br. Phycol. J.* **23**, 167–176.

Jarvis, J. D., Wynne, C. D. and Telfer, E. R. (1967). *J. Med. Lab. Technol.* **24**, 312–314.

Jones, J. B. and Stadtman, T. C. (1977). *J. Bact.* **130**, 1404–1406.

Jones, W. J., Nagle, D. P. Jr. and Whitman, W. B. (1987). *Microbiol. Rev.* **51**, 135–177.

Jong, S. C. (1978). In *The Biology and Cultivation of Edible Mushrooms*, pp. 119–135. Academic Press, London.

Joshi, L. M., Wilcoxson, R. D., Gera, S. D. and Chatterjee, S. C. (1974). *Indian J. Exp. Biol.* **12**, 598–599.

Kandler, O. (1982). *Zbl. Bact. Hyg., I. Abt. Orig. C* **3**, 149–160.

Kartha, K. K. (ed.) (1985). *Cryopreservation of Plant Cells and Organs*. CRC Press, Boca Raton.

Kartha, K. K., Leung, N. L. and Gamborg, O. L. (1979). *Plant Sci. Lett.* **15**, 7–15.

Kartha, K. K., Fowke, L. C., Leung, N. L., Caswell, K. L. and Hakman, I. (1988). *J. Plant Physiol.* **132**, 529–539.

Kilpatrick, R. A., Harmon, D. L., Loegering, W. Q., and Clark, W. A. (1971). *Plant Dis. Rep.* **55**, 871–873.

Kirsop, B. (1974). *J. Inst. Brew.* **80**, 565–570.

Kirsop, B. (1978). In *Abstracts of the XII International Congress of Microbiology, 1978*, p. 39. München.

Kirsop, B. E. and Henry, J. (1984). *Cryoletters* **5**, 191–200.

Kirsop, B. and Snell, J. J. S. (ed.) (1984). *Maintenance of Microorganisms: A Manual of Laboratory Methods*. London: Academic Press.

Korthof, G. (1932). *Zbl. Bakt. Parasitkde, I. Abt. Orig.* **125**, 429–434.

Kramer, C. L. and Mix, A. J. (1957). *Trans. Kansas Acad. Sci.* **60**, 58–64.

Langis, R. and Steponkus, P. L. (1990). *Plant Physiology* **92**, 666–671.

Langis, R., Schnabel, B., Earle, E. A. and Steponkus, P. L. (1989). *Cryoletters* **10**, 421–428.

Latham, M. J. and Sharpe, M. E. (1971). In *Isolation of Anaerobes* (D. A. Shapton and R. G. Board, eds), SAB Technical Series No. 5, pp. 134–147. Academic Press, London.

Lapage, S. P. and Redway, K. F. (1974). *Preservation of Bacteria with Notes on other Microorganisms*. Public Health Laboratory Service Monograph Series No. 7 (A. T. Willis and C. H. Collins, eds). Her Majesty's Stationery Office, London.

Lapage, S. P., Shelton, J. E., Mitchell, T. G. and MacKenzie, A. R. (1970). In *Methods in Microbiology* (J. R. Norris and D. W. Ribbons, eds), Volume 3A, pp. 135–228. Academic Press, New York.

Lapage, S. P., Redway, K. F. and Rudge, R. (1978). In *Chemical Rubber Company Handbook of Microbiology* (A. I. Laskin and H. A. Lechevalier, eds), Volume 2, pp. 743–758. Chemical Rubber Company Press, Florida.

Last, F. T., Price, D., Dye, D. W. and Hay, E. M. (1969). *Trans. Br. Mycol. Soc.* **53**, 328–330.

Larkin, P. J. and Scowcroft, W. R. (1981). *Theor. Appl. Genet.* **60**, 197–204.

Leach, C. M. (1962). *Can. J. Bot.* **40**, 151–161.

Leach, C. M. (1971). In *Methods in Microbiology*, Volume 4 (C. Booth, ed.), pp. 608–664. Academic Press, New York.

Leedale, G. F. (1967). *Euglenoid Flagellates*. Prentice-Hall, Englewood Cliff, NJ.

Leef, J. L., Strome, C. P. A. and Beaudoin, R. K. (1979). *Bull. Wld. Hlth. Org.* **57** (suppl. 1), 87–91.

Leeson, E. A., Cann, J. P. and Morris, G. J. (1984). In *The Maintenance of Microorganisms* (B. E. Kirsop and J. J. S. Snell, eds), pp. 131–160. Academic Press, London.

Loegering, W. Q. (1965). *Phytopathology* **55**, 247.

Long, R. A., Woods, J. M. and Schmitt, C. G. (1978). *Plant Dis. Rep.* **62**, 479–481.

Love, J. N. (1972). *Am. J. Vet. Res.* **33**, 2557–2560.

Love, S. L., Rhodes, B. B. and Moyer, J. W. (1987). *Practical Manuals for Handling Crop Germplasm In Vitro, 1: Meristem-tip Culture and Virus Indexing of Sweet Potato*. IBPGR, Rome.

Lumsden, W. H. R., Herbert, W. J. and McNeillage, G. J. C. (1973). *Techniques with Trypanosomes*. Churchill Livingstone, Edinburgh and London.

Luyet, B. J. and Gehenio, P. M. (1954). *J. Protozool.* **1** (suppl.), 7.

Lyman, J. R. and Marchin, G. L. (1984). *Cryobiology* **21**, 170–176.

MacDonald, K. D. (1972). *Appl. Microbiol.* **23**, 990–993.

McGarrity, C. J., Kotani, H. (1986). *Exp. Cell Res.* **163**, 273–278.

McGarrity, C. J., Kotani, H. and Carson, D. I. (1986). In Vitro *Cell Inter. Devel. Biol.* **22**, 301–304.

McGinnis, M. R., Padhye, A. A. and Ajello, L. (1974). *Appl. Microbiol.* **28**, 218–222.

McGrath, J. J. (1987). In *The Effects of Low Temperatures on Biological Systems* (G. J. Morris and B. W. W. Grout, eds), pp. 234–268. Edward Arnold, London.

McLellan, M. R. (1989). *Diatom Res.* **4**, 301–318.

McLellan, M. R., Morris, G. J. and Kalinina, L. V. (1982). *Cryoletters* **3**, 35–40.

McLellan, M. R., Schrijnemakers, E. W. M. and Van Iren, F. (1990). *Cryoletters* **11**, 189–204.

Macy, J. M., Snellen, J. E. and Hungate, R. E. (1972). *Am. J. Clin. Nutr.* **25**, 1318–1323.

Malik, K. A. (1976). In *Proceedings of Fifth International Fermentation Symposium* (H. Dellway, ed.), p. 180. Westkreuz Druckerei and Verlag, Bonn.

Malik, K. A. (1978). In *Proceeding of the XII International Congress of Microbiology*, pp. 162. *Int. Assoc. Microbiol. Soc.* München.

Malik, K. A. (1983). *J. Microbiol. Methods* **1**, 343–352.

Malik, K. A. (1984a). In *Proceedings of the Fourth International*

Conference on Culture Collection (M. Kocur and E. Da Silva eds), pp. 89–100. World Federation for Culture Collections, Brno.

Malik, K. A. (1984b). *J. Microbiol Methods* **2**, 41–47.

Malik, K. A. (1985). *Modern Methods of Gene Conservation. A Laboratory Manual.* PASTIC Press, Pakistan Science and Technology Information Centre, Islamabad, Pakistan.

Malik, K. A. (1987a). In *Stabilité et conservation des Microorganismes* (J. Amen and P. Tesson, eds), pp. 118–150. Societé Francaise de Microbiologie, Paris.

Malik, K. A. (1987b). In *Second Conference on Taxonomy and Automatic Identification of Bacteria.* M. Chyle (ed.), p. 60. Czechoslovak Soc. Microbiol. Prague.

Malik, K. A. (1987c). In *The Role of Culture Collections in the Stability and Preservation of Microorganisms* (J. Amen and P. Tesson, eds), pp. 118–150. Societé Francaise de Microbiologie, Paris.

Malik, K. A. (1988a). *J. Microbiol. Methods* **8**, 259–271.

Malik, K. A. (1988b). *J. Microbiol. Methods* **8**, 273–280.

Malik, K. A. (1989). In *Symposium on Molecular Biology of Membrane-bound Complexes in Phototrophic Bacteria.* Freiburg.

Malik, K. A. (1990a). *J. Microbiol. Methods* **12**, 117–124.

Malik, K. A. (1990b). *J. Microbiol. Methods* **12**, 125–132.

Malik, K. A. and Claus, D. (1987). In *The Biotechnology and Genetic Engineering Reviews* Volume 5 (G. E. Russel, ed.), pp. 137–197. Intercept Ltd, Ferndown, Dorset, U.K.

Margulis, L., Corliss, J. O. C., Melkonian, M. and Chapman, D. J. (1989). In *Handbook of Prototista,* pp. 914. Jones and Barlett, Boston.

Marx, D. H. and Daniel, W. J. (1976). *Can. J. Microbiol.* **22**, 338–341.

Mauer, H. R. (1986). In *Animal Cell Culture, A Practical Approach* (R. I. Freshney, ed.), pp. 13–31. IRL Press, Oxford.

Mazur, P. (1966). In *Cryobiology* (H. T. Meryman, ed.), pp. 213–315. Academic Press, New York.

Mehtra, S. K., Dayal, H. M. and Agarwal, P. N. (1977). *Indian J. Exp. Biol.* **15**, 81–82.

Meryman, H. T. (1956). *Science* **124**, 515–521.

Meryman, H. T. (1966). In *Cryobiology* (H. T. Meryman, ed.), pp. 609–663. Academic Press, New York.

Miles, A. A. and Misra, S. S. (1938). *J. Hyg. Camb.* **38**, 732–749.

Miller, T. L. (1989). In *Bergey's Manual of Systematic Bacteriology* (J. T. Staley, M. P. Bryant, N. Pfennig, and J. G. Holt, eds), Volume 3, pp. 2178–2183, 2214–2216. Williams & Wilkins, Baltimore, USA.

Miller, T. L. and Wolin, M. J. (1974). *Appl. Microbiol.* **27**, 985–987.

Mitic, S., Otenhajmer, I. and Damjanovic, V. (1974). *Cryobiology* **11**, 116–120.

Mix, G. (1985). In *Efficiency in Plant Breeding* (W. Lange, A. C. Zeven and N. G. Hogenboom, eds), pp. 194–195. Pudoc, Wageningen.

Miyata, A. (1975). *Trop. Med., Nagasaki* **17**, 55–64.

Montagnon, B. J. (1989). In *Advances in Animal Cell Biology and Technology for Bioprocesses* (R. E. Spier, J. B. Griffiths, J. Stephenson and P. J. Crooy, eds), pp. 3–15. Butterworths, London.

Morris, G. J. (1976). *Arch. Microbiol.* **107**, 309–312.

Morris, G. J. (1978). *Br. Phycol. J.* **13**, 15–24.

Morris, G. J. (1981). *Cryobiology*. Institute of Terrestrial Ecology, Cambridge.

Morris, G. J. (1987). In *The Effects of Low Temperatures on Biological Systems* (G. J. Morris and B. W. W. Grout, eds), pp. 120–146. Edward Arnold, London.

Morris, G. J., Clarke, A. and Fuller, B. J. (1980). *Cryoletters* **1**, 121–128.

Morris, G. J., Winters, L., Coulson, G. E. and Clarke, K. J. (1983a). *J. Gen. Microbiol.* **129**, 2023–2034.

Morris, G. J., Coulson, G. E., Meyer, M., McLellan, M. R., Fuller, B. J., Grout, B. W. W., Pritchard, H. W. and Knight, S. C. (1983b). *Cryoletters* **4**, 179–192.

Morris, G. J., Coulson, G. E. and Engels, M. (1986). *J. Exp. Bot.* **37**, 842–856.

Morris, G. J., Smith, D. and Coulson, G. E. (1988). *J. Gen. Microbiol.* **134**, 2897–2906.

Morris, G. J., Coulson, G. E. and Clarke, K. J. (1988). *Cryobiology* **25**, 471–482.

Mowles, J. M. (1988). *Cytotechnology* **1**, 355–358.

Muggleton, P. W. (1963). *Prog. Indust. Microbiol.* **4**, 191–214.

Mutetwa, S. (1983). Ph.D. Thesis, London University.

Mutetwa, S. M. and James, E. R. (1984a). *Cryobiology* **21**, 329–339.

Mutetwa, S. M. and James, E. R. (1984b). *Cryobiology* **21**, 552–558.

Mutetwa, S. M. and James, E. R. (1985). *Parasitology* **90**, 589–603.

Nagel, J. G. and Kunz, L. J. (1972). *Appl. Microbiol.* **23**, 837–838.

Nei, T. (1964). *Cryobiology* **1**, 87–93.

Norton, C. C. and Joyner, L. P. (1968). *Res. Vet. Sci.* **9**, 598–600.

Norton, C. C., Pout, D. D. and Joyner, L. P. (1968). *Folia Parasitol.* **15**, 203–211.

Obara, Y., Yamai, S., Nikkawa, T., Shimoda, Y. and Miyamoto, Y. (1981). *J. Clin. Microbiol.* **14**, 61–66.

Odds, F. C. (1976). *UK Federation for Culture Collections Newsletter* No. 3, December, pp. 6–7.

Ogata, W. N. (1962). *Neurospora Newsletter* **1**, 13.

Onions, A. H. S. (1971). In *Methods in Microbiology* (C. Booth, ed.), Volume 4, pp. 113–151. Academic Press, New York.

Onions, A. H. S. (1977). In *Proceedings of the Second International*

Conference on Culture Collections (A. F. Pestana de Castro, E. J. De Silva, V. B. D. Skerman and W. W. Leveritt, eds). University of Queensland, Brisbane.

Osborne, J. A. and Lee, D. (1975). *J. Protozool.* **22**, 233–236.

Otsuka, S. and Manako, K. (1961a). *Jap. J. Bact.* **16**, 814–818.

Page, F. C. (1988). *A New Key to Freshwater Gymnamoeba.* Freshwater Biological Association, Ambleside, U.K.

Palmer, D. A., Buening, G. M. and Carson, C. A. (1982). *Parasitology* **84**, 567–572.

Pautrizel, R. and Carloz, L. (1952). *C.r. Seanc. Soc. Biol.* **146**, 89.

Pearson, B. M., Jackman, P. J. H., Painting, K. A. and Morris, G. J. (1990). *Cryoletters*, in press.

Pell, P. A. and Sneath, P. H. A. (1984). *J. Appl. Bact.* **57**, 165–167.

Perkins, D. D. (1962). *Can. J. Microbiol.* **8**, 591–594.

Pfennig, N. and Trüper, H. G. (1974). In *Bergey's Manual of Determinative Bacteriology*, 8th edn (R. E. Buchanan and N. E. Gibbons, eds), pp. 24–64, Williams and Wilkins Co., Baltimore.

Pfennig, N. and Trüper, H. G. (1981). In *The Prokaryotes* (M. P. Starr, H. Stolp, H. G. Trüper, A. Balow and H. G. Schlegel, eds), pp. 279–289. Springer Verlag, Berlin.

Pfennig, N. and Trüper, H. G. (1989). In *Bergey's Manual of Systematic Bacteriology*, Volume 3 (J. T. Staley, ed.), pp. 1635–1709. Williams and Wilkins Co., Baltimore.

Phillips, B. A., Latham, M. J. and Sharpe, M. E. (1975). *J. Appl. Bact.* **38**, 319–322.

Phillips, R. E., Boreham, P. F. L. and Shepherd, R. W. (1984). *Trans. R. Soc. Trop. Med. Hyg.* **78**, 604–606.

Phillips, R. L., Kaeppler, S. M. and Peschke, V. M. (1990) In *Progress in Plant Cellular and Molecular Biology* (H. J. J. Nijkamp, L. H. W. van der Plas and J. van Aartrijk, eds), pp. 131–141. Kluwer, Dordrecht.

Polyansky, G. I. (1963). *Acta Protozool.* **1**, 166–175.

Prescott, J. M. and Kernkamp, M. F. (1971). *Plant Dis. Rep.* **55**, 695–696.

Pringsheim, E. G. (1946a). *J. Ecol.* **33**, 193–204.

Pringsheim, E. G. (1946b). *Pure Cultures of Algae.* Cambridge University Press, Cambridge.

Provasoli, L., McLauglin, J. A. and Droop, M. R. (1957). *Arch. Microbiol.* **25**, 392–428.

Raether, W. and Seidenath, H. (1972). *Tropenmed. Parasitol.* **23**, 428–431.

Raether, W. and Seidenath, H. (1977). *Z. Parasitenkd.* **53**, 41–46.

Raper, K. B. and Alexander, D. F. (1945). *Mycologia* **37**, 499–525.

Redenbaugh, K. (1990). *HortScience* **25**, 251–255.

Redway, K. F. and Lapage, S. P. (1974). *Cryobiology* **11**, 73–79.

Reed, B. M. (1989). *Cryoletters* **10**, 315–322.

Reinecke, P. and Fokkema, N. J. (1979). *Trans. Br. Mycol. Soc.* **72**, 329–331.

Reischer, H. S. (1949). *Mycologia* **41**, 177–179.

Resseler, R., Riel, J. van, and Riel, M. van (1966). *Annls. Soc. Belge Méd. Trop.* **46**, 213–22.

Rey, L. R. (1977). In *Development in Biological Standardisation.* International symposium of freeze drying of biological products, pp. 19–27, Acting editors: V. J. Cabasso and R. H. Regamey. S. Karger, Basel.

Ribeiro dos Santos, R., von Gal Furtado, C. C., Martins, J. B. and Martins, Λ. C. P. (1978). *Rev. Bras. Pesquisas Med. Biol.*, **11**, 99–104.

Roach, B. M. (1920). *Ann. Bot. Lond.* **34**, 35.

Rosenblatt, J. E., Fallon, A. and Finegold, S. M. (1973). *Appl. Microbiol.* **25**, 77–85.

Rossi, P., Pozio, E. and Besse, M. G. (1990) *Trans. R. Soc. Trop. Med. Hyg.* **84**, 68.

Rowe, T. W. G. (1971). *Cryobiology* **8**, 133–172.

Sakai, A. (1985). In *Cryopreservation of Plant Cells and Organs* (K. K. Kartha, ed.), pp. 135–158. CRC Press, Boca Raton.

Schoofs, J. (1991). In *Musa — Conservation and Documentation*, Proceedings of INIBAP-IBPGR Workshop, Leuven, 11–14 December, 1989. INIBAP/IBPGR, Montpellier. In press.

Scowcroft, W. R. (1984) *Genetic Variability in Tissue Culture: Impact on Germplasm Conservation and Utilization.* IBPGR, Rome.

Scowcroft, W. R. (1985). In *Genetic Flux in Plants* (T. Hohn and E. S. Denis, eds), pp. 217–245. Springer, Vienna.

Sharp, E. L. and Smith, F. G. (1952). *Phytopathology* **42**, 263–264.

Shearer, B. L., Zeyen, R. J. and Ooka, J. J. (1974). *Phytopathology* **64**, 163–167.

Sidyakina, T. M. (1985). *Preservation of Microorganisms* [in Russian]. Pushchino.

Sidyakina, T. M. (1988). *Methods of Microbial Preservation* [in Russian]. Pushchino.

Sidyakina, T. M. (1989). In *Proceedings of the IV International School on Cryobiology and Freeze-Drying*, 29 July to 6 August 1989, p. 155–157. Borovets, Bulgaria.

Sidyakina, T. M. and Golimbet, V. E. (1988). In *International Congress on Culture Collections*, Washington, 30 October to 4 November 1988, Abstracts, p. 75.

Sidyakina, T. M. and Golimbet, V. E. (1990). *Cryobiology*, in press.

Simione F. P. and Daggett P. M. (1976). *J. Parasit.*, **62**, 49.

Simione, F. P., Daggett, P. M., McGrath, M. S. and Alexander, M. T. (1977). *Cryobiology* **14**, 500–502.

Skerman, V. B. D. (1973). In *Proceedings of the Second International Conference on Culture Collections* (A. F. Pestana de Castro,

E. J. DaSilva, V. B. D. Skerman and W. W. Leveritt, eds), pp. 20–40. University of Queensland, Brisbane.

Sleesman, J. P., Larsen, P. O. and Safford, J. (1974). *Plant Dis. Rep.* **58**, 334–336.

Sly, L. (1988). *Maintenance and Preservation of Microbial Cultures in a Laboratory Culture Collection.* National Association of Testing Authorities, Australia, Special Technical Information Series, No. 1, April 1988.

Smith, D. (1982). *Trans. Br. Mycol. Soc.* **79**, 415–421.

Smith, D. (1983a). *Trans. Br. Mycol. Soc.* **80**, 360–363.

Smith, D. (1983b). *Trans. Br. Mycol. Soc.* **80**, 333–337.

Smith, D. (1988). In *Living Resources for Biotechnology: Filamentous Fungi* (D. L. Hawksworth and B. E. Kirsop, eds), pp. 75–99. Cambridge: Cambridge University Press.

Smith, D. and Onions, A. H. S. (1983a). *The Preservation and Maintenance of Living Fungi.* Commonwealth Mycological Institute, Kew.

Smith, D. and Onions, A. H. S. (1983b). *Trans. Br. Mycol. Soc.* **81**, 535–540.

Smith, R. S. (1967). *Mycologia* **59**, 600–609.

Smith, R. S. (1971). *Mycologia* **63**, 1218–1221.

Snyder, W. C. and Hansen, H. N. (1946). *Mycologia* **38**, 455–462.

Souzu, H. (1973). *Cryobiology* **10**, 427–431.

Staab, J. A. and Ely, J. K. (1987). *Cryobiology* **24**, 174–178.

Stackebrandt, E. and Woese, C. R. (1984). *Microbiol. Sci.* **I**, 117–122.

Staffeldt, E. E. and Sharp, E. L. (1954). *Phytopathology* **44**, 213–214.

Stalpers, J. A., DeHoog, A. and Vlug, I. J. (1987). *Mycologia* **79**, 82–89.

Stamp, L. (1947). *J. Gen. Microbiol.* **1**, 251–265.

Stetter, K. O., Thomm, M., Winter, J., Wildgruber, G., Huber, H., Zillig, W., Janecovic, D., König, H., Palm, P. and Wunderl, S. (1981). *Zbl. Bact. Hyg., I. Abt. Orig. C* **2**, 166–178.

Swaroop, S. (1938). *Indian J. Med. Res.* **26**, 353–378.

Takeuchi, M., Matsushima, H. and Sugawara, Y. (1982). In *Plant Tissue Culture, 1982.* (A. Fujiwara, ed.), pp. 797–798. Maruzen, Tokyo.

Tatewaki, M., Provasoli, L. and Printer, I. J. (1983). *J. Phycol.* **19**, 409–416.

Thompson, A. S., Rhodes, J. C. and Pettman, I. (1988). *Culture Collection of Algae and Protozoa: Catalogue of Strains.* CCAP, Windermere.

Timnick, M. B., Lilly, V. G. and Barnett, H. L. (1951). *Phytopathology* **41**, 327–336.

Towill, L. E. (1985). In *Cryopreservation of Plant Cells and Organs* (K. K. Kartha, ed.), pp. 171–198. CRC Press, Boca Raton.

Tuite, J. (1968). *Mycologia* **60**, 591–594.

Trüper, H. G. and Imhoff, J. F. (1981). In *The Prokaryotes* (M. P. Starr, H. Stolp, H. G. Trüper, A. Balow and H. G. Schlegel, eds), pp. 274–278. Springer Verlag, Berlin.

Turner, L. H. (1970). *Trans. R. Soc. Trop. Med. Hyg.* **64**, 624–646.

Turner, M. F. (1979). *J. Mar. Ass. U.K.* **56**, 535–552.

Turner, M. F. and Droop, M. R. (1978). In *CRC Handbook Series in Nutrition and Food* (M. Riecheigl Jr., ed.), Section G **3**, pp. 287–426.

Umebeyashi, O. (1972). *Bull. Tokai Region. Fish. Res. Lab.* **69**, 55–62.

Uragami, A., Sakai, A., Nagai, M. and Takahashi, T. (1989). *Plant Cell Rep.* **8**, 418–421.

USSR Author's Certificate No. 1442544. Bull. Izobreteniy, N 45 (1988).

Van Niel, C. B. (1971). In *Methods in Enzymology*, Volume **23**, Part A (A. San Pietro, ed.), pp. 3–28, Academic Press, New York.

Vega, C. A., Buening, G. M., Rodriguez, S. D., Carson, C. A. and McLaughlin, K. (1985). *Am. J. Vet. Res.* **46**, 421–423.

Volkman, J. K., Jeffrey, S. W., Nichols, P. D., Rodgers, G. I. and Garland, C. D. (1989), *J. Exp. Mar. Biol. Ecol.* **128**, 219–240.

von Arx, J. A. and Schipper, M. A. A. (1978). *Adv. Appl. Microbiol.* **24**, 215–236.

Vuylsteke, D., Swennen, R., Wilson, G. F. and De Langhe, E. A. (1988). *Scientia Horticulturae* **36**, 70–88.

Vuylsteke, D. and Swennen, R. (1990) *IITA Research* **1**(1), 470.

Walker, P. J. (1966). *Lab. Pract.* **15**, 423–426.

Wanas, W. H., Callow, J. A. and Withers, Lyndsey A. (1986). In *Plant Tissue Culture and its Agricultural Applications* (L. A. Withers and P. G. Alderson, eds), pp. 285–289. Butterworths, London.

Watanabe, B., Yamada, Y., Ueno, S. and Mitsuda, H. (1985). *Agric. Biol. Chem.* **49**, 1727–1731.

Webber, W. A. F., Cunningham, M. P. and Lumsden, W. H. R. (1961). *East African Trypanosomiasis Research Organization Annual Report, January–December 1961*, 10–11.

Webster, J. and Davey, R. A. (1976). *Trans. Br. Mycol. Soc.* **67**, 543–544.

Wellman, A. M. (1971). *Cryobiology* **7**, 259–262.

Wellman, A. M. and Stewart, G. G. (1973). *Appl. Microbiol.* **26**, 577–583.

Whitman, W. B. (1989). In *Bergey's Manual of Systematic Bacteriology* (J. T. Staley, M. P. Bryant, N. Pfennig and J. G. Holt, eds), Volume **3**, pp. 2185–2190. Williams & Wilkins, Baltimore.

Whittingham, D. G., Wood, M., Farrant, J., Lee, H. and Halsey, J. A. (1979). *J. Reprod. Fert.* **56**, 11–21.

Wildgruber, G., Thomm, M., König, H., Ober, K., Ricchiuto, T. and Stetter, K. O. (1982). *Arch. Microbiol.* **132**, 31–36.

Williams, D. L. and Calcott, P. H. (1982). *J. Gen. Microbiol.* **128**, 215–218.

Wilson, R. J. M., Farrant, J. and Walter, C. A. (1977). *Bull. Wld. Hlth. Org.* **55**, 309–315.

Winter, J. (1983). *System. Appl. Microbiol.* **4**, 558–563.

Withers, L. A. (1979). *Plant Physiol.* **63**, 460–467.

Withers, L. A. (1980a). *Tissue Culture Storage for Genetic Conservation.* IBPGR, Rome.

Withers, L. A. (1980b). *Cryoletters* **1**, 239–250.

Withers, L. A. (1982). In *Crop Genetic Resources — The Conservation of Difficult Material* (L. A. Withers and J. T. Williams, eds), pp. 49–82. IUBS Series B42. IUBS/IBPGR/IGF, Paris.

Withers, L. A. (1984). In *Cell Culture and Somatic Cell Genetics of Plants. Vol. 1: Laboratory Procedures and Their Applications* (I. K. Vasil, ed.), pp. 608–620. Academic Press, Orlando.

Withers, L. A. (1985a). In *Cell Culture and Somatic Cell Genetics of Plants. Vol. 2: Cell Growth, Nutrition, Cytodifferentiation and Cryopreservation* (I. K. Vasil, ed.) pp. 253–316. Academic Press, Orlando.

Withers, L. A. (1985b). In *Cryopreservation of Plant Cells and Organs* (K. K. Kartha, ed.), pp. 243–267. CRC Press, Boca Raton.

Withers, L. A. (1985c). In *Plant Cell Culture: A Practical Approach* (R. A. Dixon, ed.), pp. 169–191. IRL Press, Oxford.

Withers, L. A. (1986). In *Plant Cell Culture Technology* (M. M. Yeoman, ed.), pp. 96–140. Blackwell, Oxford.

Withers, L. A. (1987a). *Ox. Surv. Plant Mol. Cell Biol.* **4**, 221–272.

Withers, L. A. (1987b). In *The Effects of Low Temperatures on Biological Systems* (B. W. W. Grout and J. G. Morris, eds), pp. 389–409. Edward Arnold, London.

Withers, L. A. (1989). In *The Use of Plant Genetic Resources* (A. D. H. Brown, D. R. Marshall, O. H. Frankel and J. T. Williams, eds), pp. 309–334. Cambridge University Press, Cambridge.

Withers, L. A. (1990). In *Methods in Molecular Biology*, Volume 5 (J. M. Walker and J. W. Pollard, eds), pp. 39–48. Humana Press, New Jersey.

Withers, L. A. (1991a). In *Proceedings of National Conference on Plant and Animal Biotechnology*, Nairobi, Kenya, 25 February–3 March 1990. In press.

Withers, L. A. (1991b). In *Proceedings of International Workshop on Tissue Culture for the Conservation of Biodiversity and Plant Genetic Resources*, Kuala Lumpur, 28–31 May 1990. In press.

Withers, L. A. and King, P. J. (1980). *Cryoletters* **1**, 213–220.

Withers, L. A. and Williams, J. T. (eds) (1982). *Crop Genetic Resources — The Conservation of Difficult Material.* IUBS Series B42. IUBS/IBPGR/IGF, Paris.

Withers, L. A., Benson, Erica E. and Martin, M. (1988). *Cryoletters* **9**, 114–119.

Wolff, J. W. (1960). In *The Leptospirae and Leptospirosis in Man and Animals*. The Preblem Session Series of the Polish Academy of Sciences **19**, 11–15.

Wood, D. J. and Minor, P. D. (1990). *Biologicals* **18**, 143–146.

Woods, R. (1976). *UK Federation for Culture Collections Newsletter* No. **2**, p. 5.

Yaeger, R. G. (1988). *J. Protozool.* **35**, 114–115.

Yamai, S., Obara, Y., Nikkawa, T., Shimoda, Y. and Miyamoto, Y. (1979). *Br. J. Vener. Dis.* **55**, 90–93.

Zehnder, A. J. B. and Wuhrmann, K. (1977). *Arch. Microbiol.* **111**, 199–205.

Zeikus, J. G. (1977). *Bact. Rev.* **41**, 514–541.

Zeikus, J. G. and Henning, D. L. (1975). *Antonie von Leeuwenhoek Microbiol. Serol.* **41**, 543–552.

Zeikus, J. G. and Wolfe, R. S. (1972). *J. Bact.* **109**, 707–713.

APPENDIX I

DOCUMENTATION

An efficient documentation system is an essential accompaniment to culture maintenance and the following lists provide guidance on the kind of information that should be recorded. Details of information required will vary between laboratories. Data may be stored in notebooks, files, card indexes, punch cards, or in semi- or fully-computerized systems. A number of suitable computer programs exist and advice on appropriate selection can be obtained from any of the culture collections or data centres listed in Chapter 3.

A. Strain records

1. Essential
a. laboratory reference number
b. alternative reference numbers (service collections, other laboratories)
c. name, if known
d. date of deposit
e. cultural requirements (medium, temperature, pH, nutritional requirements . . .)
f. maintenance method (subculture at 4-week intervals/freeze-drying using *'mist-desiccans'*/2-stage freezing in 10% glycerol . . .)

2. Desirable
a. use of culture (production strain, assay strain, sterility testing . . .)
b. cultural properties (physiological, morphological, genetical . . .)
c. references

B. Maintenance records

1. *sub-culturing information*: medium, temperature,
storage period records: date, cultural characteristics, special notes . . .
2. *freeze-drying information*: method of growth and cell concentration of inoculum, suspending medium, numbers of ampoules to be prepared, storage and revival procedures . . .
records: date of processing, batch number, viability counts, date for reprocessing, number of ampoules in storage . . .
3. *liquid nitrogen storage information*: method of growth and cell concentration of inoculum, cryoprotectant, freezing rate, number of ampoules/capillaries/straws to be prepared, storage and revival procedures . . .
records: date of freezing, batch number, viability counts, date of reprocessing, number of ampoules/capillaries/straws in storage . . .

C. 'In-house' records for supply of cultures

Date, person requesting culture, purpose of request, feedback on culture performance . . .

Example:

Strain no.	Date requested	Requested by	Purpose	Feedback
XYZ83	6.4.83	BHK	pilot plant run 41	normal performance

LIST OF SUPPLIERS

Adelphi Manufacturing, 20–21 Duncan Terrace, London, N1 8BZ, UK.

L' Air Liquide, 57, Avenue Carnot, F-94500 Champigny-s/Marne, France

Anchor Glass Co. Ltd, Brent Cross Works, North Circular Road, London, NW4 1JS, UK.

Armour Pharmaceuticals Co. Ltd, Hampton Park, Eastbourne, Sussex, BN21 3YG, UK.

Baird and Tatlock (London) Ltd, P.O. Box 1, Freshwater Road, Chadwell Heath, Romford, Essex, RM1 1HA, UK.

BDH Chemicals Ltd, Poole, Dorset, BH12 4NN, UK.

B-D Laboratory Products Division, Becton Dickinson UK Ltd, York House, Empire Way, Wembley, Middx. HA9 0PS, UK.

Bellco Glass Inc., Vineland, NJ, USA.

Peter Bellingham Ltd, Unit 5/6, Chailey Industrial Estate, Pump Lane, Hayes, Middlesex, UB3 3NB, UK.

Bibby NUNC, NUNC Intermediates, Kamstrup, DK-4000, Denmark.

Boots Pure Drug Co. Ltd, Lenton Research Station, Lenton House, Nottingham, NG7 2QD, UK.

Rudolph Brand, P.O. Box 310, D-6980 Wertheim, FRG.

British Drug Houses Ltd, Laboratory Chemicals Division, Poole, Dorset, UK.

Buck & Hickman, Sterling Industrial Estate, Rainham Road South, Dagenham, Essex, RM10 8TA, UK.

Business Simulations Ltd, 30 St. James Street, London, SW14 1HB, UK.

(Caelo) Caesar & Loretz GmbH, Herderstrasse 31 or Herderstr. 31, D-4010 Hilden, FRG.

Camlab Ltd, Nuffield Road, Cambridge, CB4 1TH, UK.

Creative Beadcraft Ltd, Denmark Works, Sheepcote Dell Road, Beamond, Amersham, Bucks, HP7 0RX, UK.

Cryo-Med, 51529 Birch Street, New Baltimore, MI 48047, USA.

Denley Instruments Ltd, Billingshurst, Sussex, RH14 9EY, UK.

Difco Laboratories, P.O. Box 14B, Central Avenue, East Molesey, Surrey, KT8 0SE, UK.

Domino Printers Ltd, Bar Hill, Cambridge, CB3 8TU, UK.

Dragonas Microflame, Vinces Road, Diss, Norfolk, UK.

Edwards High Vacuum International, Manor Royal, Crawley, West Sussex, RH10 2LW, UK.

Ells and Farrier Ltd., 5 Prince's Street, Hanover Square, London, W1R 8PH, UK.

FBG-Trident Ltd, Temple Cloud, Bristol, BS18 5BY, UK.

Fryka Kaltetechnic, (UK Supplier), Comlab Ltd., Nuffield Road, Cambridge, CB4 1TH, UK.

Gallen Kamp, Belton Road West, Loughborough, Leicestershire, LG11 0TR, UK.

Gelman Sciences Ltd, Brackmills Business Park, Caswell Road, Northampton, NN4 0EZ, UK.

Genzyme Biochemicals Ltd, Springfield Mill, Maidstone, Kent, ME14 2LE, UK. *Distributors:* Scientific Supplies Co. Ltd, Scientific House, Vine Hill, London, EC1R 5EB, UK.

Glass Wholesale Supplies Ltd, 566 Cable Street, London, EW1 9EZ, UK.

Goldschmidt Ltd, Chemical Products, Tego House, Victoria Road, Ruislip, Middlesex, HA4 0YL, UK.

Harris Manufacturing Co. Inc., Billerico, Mass., USA.

Philip Harris Scientific, 618 Western Avenue, Park Royal, London, W3 0TE, UK.

Harshaw Chemicals Ltd, P.O. Box 4, Daventry, Northants, UK.

IMV Ltd, IOBD Clemenceau BP 76, 61300 L'Aigle, France.

ICI Agrochemicals UK Sales, Woolmead House West, Bear Lane, Farnham, Surrey, GU9 7UB, UK.

ICN Flow Ltd, P.O. Box 17, Second Avenue Industrial Estate, Irvine, Ayrshire, KA12 8NB, UK.

Imperial Laboratories Ltd, West Portway, Andover, Hampshire, SP10 3LF, UK.

Instruments de Médecine Vétérinaire, IOBD Clemenceau BP 76, 61300 L'Aigle, France.

International Marketing Corporation, 36 Lenexa Business Center, 9900 Pflumm Road, Lenexa, Kansas 66215, USA.

Jencons (Scientific) Ltd, Cherrycourt Way Industrial Estate, Stanbridge Road, Leighton Buzzard, Bedfordshire, LU7 8UA, UK.

Koch-Light Ltd, 163 Dixons Hill Road, North Mimms, Hatfield, Hertfordshire, AL9 7JE, UK.

Lab-Lemco (as Unipath Ltd)

Lab M, Topley House, P.O. Box 19, Bury, Lancashire, BL9 6AU, UK.

Laboratory Environmental Supply Associates, 2 Elmside, Green Lane, Hardwicke, Gloucester, GL2 6OF, UK.

Life Technologies Ltd (Gibco), P.O. Box 35, Trident House, Kenfrew Road, Paisley, PA3 4EF, UK.

London Analytical & Bacteriological Media Ltd, Ford Lane, Salford, M6 6PB, UK.

Longs Ltd, Hanworth Trading Estate, Chertsey, Surrey, KT16 9LZ, UK.

Marion Scientific Corporation, Kansas City, MI, USA.

E. Merck, Postfach 4119, D-1600 Darmstadt, FRG.

Messer Griesheim GmbH, Postfach 47 09, D-4000 Düsseldorf, FRG.

Mi-Dox Ltd, Smarden, Kent, TN27 8OL, UK.

Millipore (UK) Ltd, Millipore House, 11–15 Peterborough Road, Harrow, Middx., HA1 2YH, UK.

Murphy Chemical Co., Wheathampstead, St. Albans, Herts, AL4 8QY, UK.

Nucleopore Corporation, 7035 Commerce Circle, Pleasanton, CA 94566, USA.

A/S Nunc, Postbox 280, Kastrup, DK-4000, Denmark.

Patterson Scientific, Unit 2, Brookside, Colne Way, Watford, Hertfordshire, WD2 4QJ, UK.

Perrier UK Ltd, 6 Lygon Place, London, SW1W 0JR, UK.

Planer Products, Windmill Road, Sunbury-on-Thames, Middlesex, TW16 7HD, UK.

Plowden and Thompson Ltd, Dial Glass Works, Stourbridge, West Midlands, DY8 4YN, UK.

John Poulten Ltd, 77–93 Tanner Street, Barking, Essex, IG11 8OD, UK.

Raven Scientific Ltd, P.O. Box 2, Haverhill, Suffolk, CB9 7UU, UK.

Rejafix Ltd, Harlequin Avenue, Great West Road, Brentford, Middlesex, TW8 9EH, UK.

Scientific Supplies Co. Ltd, Scientific House, Vine Hill, London, EC1R 5EB, UK.

Scotts Office Equipment Ltd, Deseronto Wharf, St. Mary's Road, Langley, Slough, Berks, SL3 7EW, UK.

Sigma Chemical Company, P.O. Box 14508, St Louis, MO 63178, USA.

Sterilin Instruments, 43–45 Broad Street, Teddington, Middlesex, TW11 8QZ, UK. *Distributors:* R. and L. Slaughter Ltd, 162 Baigores Lane, Gidea Park, Romford, Essex, RM2 6BS, UK.

Statebourne (Cryogenics) Ltd, Paison Industrial Estate, Washington, Tyne and Wear, NE37 1EZ, UK.

Sweetheart International Ltd, Rowner Road, Gosport, Hampshire, PO13 0PR, UK.

Techne Cambridge Ltd, Duxford, Cambridge, CB2 4PZ, UK.

Tissue Culture Services Ltd, Botolph Claydon, Bucks, MK18 2LR, UK.

Unipath Ltd, Wade Road, Basingstoke, Hampshire, RG24 0PW, UK.

Union Carbide, Cryogenics Division, Redworth Way, Aycliffe Industrial Estate, Co. Durham, DL5 6HE, UK. *Distributors:* Jencons (Scientific) Ltd, Cherrycourt Way Industrial Estate, Stanbridge Road, Leighton Buzzard, Beds, LU7 8UA, UK.

Universal Stationers, Lonsdale House, Empire Way, Wembley, Middx., HA9 0XN, UK.

Varian GmbH, Alsfelderstrasse 6 or Alsfelderstr. 6, D-6100 Darmstadt, FRG.

Waterlife Research Industries Ltd, 476 Bath Road, Longford, West Drayton, Middlesex, UB7 0ED, UK.

Wellcome Reagents Ltd, Wellcome Research Laboratories, Beckenham, BR3 3BS, UK.

Werner Aquatech, 6423 Wartenberg, FRG.

Wesley Coe (Wingent) Ltd, 115–117 Cambridge Road, Milton, Cambridge, CB4 4AY, UK.

Wheaton Inc., 1000N Tenth Street, Millville, New Jersey, NJ 08332, USA.

Don Whitley Scientific Ltd, Green Lane, Baildon, Shipley, West Yorks, BD17 85JS, UK.

T. W. Wingent, 115–150 Cambridge Road, Milton, Cambridge, UK.

A. D. Wood (London) Ltd, Service House, 1 Lansdowne Road, London, N17 0LH, UK.

GENUS INDEX

Absidia, 153
Acanthamoeba, 198, 200
Acanthocystis, 200
Achromobacter, 39
Achyla, 142, 152, 156, 157
Acinetobacter, 39
Acremoniella, 145
Acremonium, 145, 153, 154
Actinobacillus, 39, 50
Actinomadura, 39
Actinomucor, 153
Actinomyces, 39
Actinophrys, 200
Actinosphaerium, 201
Aegyptianella, 213
Aerococcus, 39
Aeromonas, 39
Alcaligenes, 39, 132
Allomyces, 142, 147, 152
Alternaria, 145, 153, 156
Alteromonas, 39
Alysiella, 39
Amoeba, 200, 204
Amorphotheca, 153
Anaerobiospirillum, 39
Anaplasma, 213
Aquaspirillum, 38, 39, 132
Arcella, 200
Areolospora, 152
Armillaria, 147, 152
Arnium, 147
Arthrobotrys, 145, 147
Arthrocladium, 152
Arthroderma, 153
Aschersonia, 156
Ascocalvatia, 152
Ascochyla, 147
Ascocoryne, 154
Ascomycotina, 147, 156, 157
Ascotricha, 147, 153
Aspergillus, 143, 148, 153, 154
Assulina, 200
Astasia, 198, 199

Aureobasidium, 153, 156
Azomonas, 60
Azospirillum, 132
Azotobacter, 132

Babesia, 213
Bacillus, 39
Bacterionema, 39
Bacteroides, 38, 39, 50, 55
Balansia, 152
Basidiobolus, 145, 148, 155
Basidiomycotina, 147, 156, 157
Batteraea, 152
Beltrania, 145, 148
Beltraniella, 145
Beneckea, 39
Bifidobacterium, 39
Biscogniauxia, 152
Bispora, 153
Blepharisma, 204
Boletus, 142
Bordetella, 39
Borrelia, 132
Botryodiplodia, 148
Botryosphaeria, 143
Botryotrichum, 153
Brachyspira, 39
Branhamella, 39
Brettanomyces, 166, 174, 177
Brevibacterium, 40
Brochothrix, 40
Brucella, 40
Bullera, 166, 174, 177
Buttiauxella, 40
Butyribacterium, 40
Byssochlamys, 153

Calonectria, 145, 148, 151
Camposporium, 152
Campylobacter, 40, 132
Candida, 148, 168, 171, 174, 177
Capnocytophaga, 40

Cardiobacterium, 40
Cedecea, 40
Cellulomonas, 40
Cephalophora, 153
Cephalosporium, 153
Ceratocystis, 143, 148
Cercospora, 145, 148, 153
Chaetodiplopodia, 153
Chaetomidium, 148, 153
Chaetomium, 153, 154
Chaetopsina, 153
Chaetosphaeria, 145
Chalara, 153
Chalaropsis, 145
Chaos, 200
Chilomonas, 198, 199, 200
Chlamydomyces, 143
Chlamydomonas, 132, 181, 204
Chorella, 194, 202, 206
Chloridium, 153
Chlorobium, 89, 96, 132
Chloroflexus, 82, 87, 89, 96, 132
Chlorogonium, 200
Chloroherpeton, 82
Chloronema, 82
Chromobacterium, 40
Chryseomonas, 40
Chrysosporium, 145
Chytridium, 148, 152
Circinella, 153
Citeromyces, 177
Citrobacter, 40
Cladobotryum, 152
Cladosporium, 153
Claviceps, 145
Clostridium, 38, 40, 50
Cochliobolus, 148
Coemansia, 148
Cokeromyces, 135
Colletotrichum, 145, 148
Colpidium, 200, 201
Comamonas, 40
Conidiobolus, 148
Coniella, 153
Coniothyrium, 153
Coprinus, 142, 148
Cordyceps, 153

Coriolus, 152
Corticium, 142, 143, 148
Cortinarius, 142
Coryne, 153
Corynebacterium, 40, 50, 55
Corythion, 200
Cryptococcus, 165, 166, 168, 177
Cryptospora, 145
Cryptosporidium, 213
Culicinomyces, 148
Cunninghamella, 153
Curvularia, 153, 154
Cyclotella, 194, 195
Cylindrocarpon, 145, 151, 153, 154
Cylindrocladium, 151
Cylindrocystis, 203
Cytophaga, 40

Dactuliophora, 152
Dekkera, 166, 174
Dermatophilus, 40
Deuteromycotina, 147, 156, 157
Diaporthe, 157
Dictyostelium, 135, 148
Didymella, 153
Didymosphaeria, 153
Dientamoeba, 214
Difflugia, 200
Distigma, 199
Doratomyces, 153
Drechslera, 143, 153
Dunaliella, 132, 194

Echinosporangium, 145
Edwardsiella, 40
Eikenella, 40
Eimeria, 214
Eladia, 153
Eleutherascus, 143, 152
Elsinoe, 153
Embellisia, 154
Endophragmia, 145
Entamoeba, 211, 214, 223
Enterobacter, 40, 50
Entomophthora, 152

Eremomyces, 152
Eremothecium, 153
Erwinia, 40
Erysipelothrix, 41
Erythrobacter, 82
Escherichia, 41, 69, 181
Eubacterium, 41, 74
Euglena, 190, 198
Euglypha, 200
Eumycota, 147
Ewingella, 41
Exobasidiellum, 148

Flavobacterium, 41
Fomes, 152
Francisella, 41
Fusarium, 148, 151, 153, 154
Fusobacterium, 41

Gaeumannomyces, 148
Ganoderma, 152
Gardnerella, 41
Gelasinospora, 148, 153
Gemella, 41
Geomyces, 153
Georgefischeria, 145
Geosmithia, 148, 153
Geotrichum, 153
Giardia, 214
Gibberella, 151
Gibertella, 153
Gliocladium, 153
Gliomastix, 153
Gloeosporium, 145
Glomerella, 145, 153
Gnomonia, 148
Gyropaigne, 199

Haematococcus, 194, 206
Haemophilus, 38, 41, 50, 55
Hafnia, 41
Hanseniaspora, 166
Hansenula, 177
Hartmanella, 204, 205

Helerocephalum, 149
Helicobacter, 38
Helicodendron, 143, 148
Helicosporina, 145, 149
Heliobacterium, 82, 88, 96, 132
Heliothrix, 82
Heliscus, 145, 149
Helminthosporium (*Drechslera*), 143, 146
Herpotrichia, 152
Hirsutella, 153
Humicola, 143, 153
Hyalophacus, 199
Hydrogenophaga, 132
Hypomyces, 145
Hypoxylon, 149

Ipomoea, 250
Isoachyla, 142

Khawkinea, 199
Khuskia, 145
Kingella, 41
Klebsiella, 41
Kloeckera, 166, 177
Kluyvera, 41, 132
Kluyveromyces, 177
Koserella, 41
Kretschmaria, 152
Kurthia, 41

Lacellinopsis, 152
Lactobacillus, 41, 50
Lasiobolidium, 152
Legionella, 41
Leishmania, 212, 214
Leminorella, 41
Lentinus, 152, 156, 157
Lenzites, 152
Leptographium, 153
Leptoporus, 152
Leptosphaeria, 149
Leptospira, 115–120
Leptotrichia, 38
Leuconostoc, 41

Leucosporidium, 132, 174
Levinea, 41
Lipomyces, 166, 177
Listeria, 42
Lomachashaka, 152
Lophiostroma, 145
Loramyces, 153

Mammaria, 153
Manihot, 249, 250
Marasmius, 152
Mariannaea, 153
Martensiomyces, 149, 152
Mastigomycotina, 147, 156, 157
Megasphaera, 74
Melanospora, 145, 151
Memmoniella, 153
Menoidium, 199
Mesobotrys, 145
Metarhizium, 149
Methanobacterium, 103, 111
Methanobolus, 105
Methanobrevibacter, 103, 111
Methanococcoides, 105
Methanococcus, 103, 104, 111
Methanohalophilus, 105, 111
Methanoplanus, 106, 111
Methanosarcina, 103, 105, 111
Methanosphaera, 104
Methanospirillum, 111
Methanothermus, 103, 106, 111
Metschnikowia, 177
Microascus, 153
Micrococcus, 42, 50
Micromonospora, 149
Mitsuokella, 42
Mobiluncus, 42
Moellerella, 42
Monacrosporium, 145, 149
Monascus, 145
Monodictys, 153
Monostroma, 185
Monotosporella, 152
Moraxella, 42
Morococcus, 42
Mortierella, 143, 145, 149, 156, 157

Mucor, 143, 153, 156, 157
Musa, 247, 250
Mycena, 142
Mycobacterium, 38, 42
Mycococcus, 42
Mycosphaerella, 149, 153
Mycovellosiella, 145, 149
Myrothecium, 153
Myxomycota, 147

Naegleria, 198, 200, 215
Nectria, 144, 145, 151, 153, 154
Neisseria, 38, 42, 50, 55, 132
Nematospora, 177
Neocosmospora, 153
Neodeightonia, 153
Netzelia, 200
Neurospora, 149, 153
Nocardia, 42
Nodularia, 206
Nodulisporium, 145
Normuraea, 149
Nostoc, 206
Nummularia, 152

Oerskovia, 42
Oidiodendron, 153
Oligella, 42
Ophiobolus, 153
Ophiovalsa, 149

Pachysolen, 177
Paecilomyces, 153, 154
Panus, 152
Paramecium, 198
Paratetramitus, 200
Parauronema, 198
Parmidium, 199
Pasteurella, 42, 50
Pediococcus, 42
Penicillifer, 145, 152
Penicillium, 135, 142, 144, 149, 153, 154, 156, 157
Peptococcus, 38, 42

Peptostreptococcus, 42
Peranema, 198
Periconia, 145
Pestalotiopsis, 153, 154
Petriella, 144
Petriellidium, 153
Phaeoisariopsis, 152
Phaeotrichoconis, 153, 154
Phaffia, 177
Phialocephala, 153
Phialomyces, 153, 154
Phialophora, 153
Phoma, 149, 153
Phomopsis, 153, 154
Phycomyces, 149, 153
Phyllosticta, 152
Physarum, 152
Phytophthora, 132, 142, 144, 149, 152,
    155, 156, 157
Pichia, 177
Piedraia, 152
Piptocephalis, 135, 142, 149
Pirella, 153
Pithomyces, 153
Plasmodium, 211, 215, 223, 225
Platystomum, 152
Pleospora, 149
Plesiomonas, 42
Pleurophragmium, 153
Podospora, 144
Polychaos, 200
Polytoma, 198
Polytomella, 198
Propionibacterium, 42, 50
Proteus, 42
Providencia, 42
Prunus, 250
Pseudocercospora, 145
Pseudomonas, 43, 55, 63, 132, 181
Puccinia, 151, 152
Pycnoporus, 154
Pyrenochaeta, 145
Pyrenophora, 145, 149
Pyrenopeziza, 149
Pyronema, 141, 149
Pyrus, 250
Pythium, 132, 142, 150, 151, 152, 156, 157

Quarternaria, 152

Ramibacterium, 43
Raphidiophrys, 201
Rhabdomonas, 199
Rhinotrichum, 153
Rhizoctonia, 144, 150
Rhizophydium, 150, 155
Rhizopus, 150, 153
Rhodobacter, 94
Rhodococcus, 43
Rhodocyclus, 94
Rhodomicrobium, 84
Rhodopseudomonas, 84, 94
Rhodospirillum, 84, 94, 96
Rhodosporidium, 174, 177
Rhodotorula, 150, 165, 166, 168, 177
Rothia, 43
Ryparobius, 150

Saccharomyces, 132, 153, 168, 169, 171,
    177, 180
Saccharomycoides, 177
Sagenomella, 154
Salmonella, 43
Saprolegnia, 142, 150, 152, 156, 157
Sarcina, 43
Sassaurea, 250
Schizophyllum, 156, 157
Schizosaccharomyces, 166, 168, 169, 177
Sclerospora, 155
Sclerotinia, 150
Sclerotium , 144, 150
Scoiulariopsis, 144, 153, 154
Searchomyces, 152
Seiridium, 150
Selenosporella, 152
Selinia, 152
Serpula, 150, 155, 156, 157
Serratia, 43
Sesquicillium, 153
Setosphaeria, 144, 153
Shigella, 43
Sigmoidea, 152
Simonsiella, 43

Solanum, 249, 250
Sordaria, 156, 157
Spegazzinia, 153
Spermospora, 145
Sphaerobolus, 150, 152
Spirillum, 60, 132
Spirulina, 194
Spondylocladiopsis, 152
Sporendonema, 144
Sporidesmium, 154
Sporobolomyces, 150, 153, 156, 157, 165,
    166, 174, 177
Sporophora, 153
Sporothrix, 153
Stachybotryna, 150
Stachybotrys, 153
Staphylococcus, 43, 50, 63
Staphylotrichum, 153
Stemphylium, 153
Stephanosporium, 144
Stereum, 152
Stigmina, 145
Stilbella, 153
Stilbum, 150
Streptobacillus, 43
Streptococcus, 43, 50
Streptomyces, 43, 150, 153
Sydowia, 153
Sympodiella, 152
Syncephalastrum, 153
Syzygites, 150, 152

Tetracladium, 152
Tetrahymena, 198, 199, 200, 201, 205
Tetranacrium, 152
Tetraselmis, 190, 194
Thamnostylum, 153
Thanatephorus, 144
Thermactinomyces, 43
Thermoascus, 153
Thielavia, 150, 151, 153
Thielaviopsis, 144
Torula, 144
Torulaspora, 177
Toxoplasma, 216

Trichoderma, 150, 153, 156, 157
Trichomonas, 216
Trichophaea, 153
Trichophyton, 156, 157
Trichothecium, 150
Trigonopsis, 168, 177
Trinema, 200
Tritirachium, 150
Trypanosoma, 217, 223

Ulocladium, 153
Unbelopsis, 152
Urohendersonia, 152
Uronema, 198
Ustilaginoidea, 152
Ustilago, 144
Ustulina, 152

Valsa, 153
Veillonella, 43, 74
Venturia, 153
Verticillium, 144
Vibrio, 43, 50, 132
Volutella, 144, 153
Volvariella, 152, 155, 156, 157

Wallemia, 156, 157
Weeksella, 43
Wingea, 177
Wolinella, 43

Yarrowia, 169, 177
Yersinia, 43, 50

Zalerion, 150
Zooglea, 43
Zopfiella, 150
Zygomycotina, 147, 156, 157
Zygorhynchus, 153
Zygosaccharomyces, 177
Zymophilus, 132

SUBJECT INDEX

Acaricides, 158–159
Acteltic, 158
Activated charcoal, protection against photo-
oxidation, 94, 96, 100, 127, 129
Agar slopes
anaerobic bacteria, cryopreservation on MM10
agar, 76
fungi
cryopreservation spore suspension, 139
storage and survival, 152–153
yeasts, 165–166
Age of culture
for preservation, general principles, 130
for preservation *see under* specific organisms
Algae
age of culture for preservation, 191, 202
air-drying, 206
for aquaculture, 194
axenic media, 185, 186, 191, 193, 194, 198
cryopreservation, 201–206
cold shock, 204
cooling, 201–205
cryoprotectants, 202
thawing, 202–203, 206
two step, 203
culture vessels, 191
freeze-drying, 206–207, 207
industrial scale production, 194
light regimes, 192
media, 186–192
shelf life
air-drying, 206
drying, 206, 207
freeze-drying, 207
stability, 193
storage, 184
strain establishment, 185
subculturing, 185–195
survival, 204
temperature regimes, 192–193
thawing, 202–203, 205, 206
viability, 205
Amoebae, media, 199–200
Ampoules, glass
general notes and cautions, 37–38
opening, 35, 79
Ampoules, polypropylene
advantages, 218, 220
alternative to glass, 155, 237
Anaerobic bacteria *see* Bacteria
Anaerobic jars, 72, 105, 113

Animal cells
adherent cell lines, 232
continuous cell lines, 228
cooling, 238
cryopreservation, 225, 236–239
thawing, 238
facilities for handling, 230
finite cell lines, 228
growth media, 229–230
hybridomas, 229
primary cell cultures, 228
quality control
detection of contamination, 239
species and function verification, 239
reconstitution and rehydration, 238
subculturing
adherent cell lines, 232–234
suspension cell lines, 234
viable cell counts, 234–236
thawing, precautions, 238
transformed cell lines, 228
viability, 234
appendices, 287–292
Aquaculture, 194
Archaebacteria *see* Methanogenic bacteria
ASP2 marine mineral medium, 188
Axenic media, 185, 186, 191, 193, 194, 198

Bacteria
age of culture for preservation, 32, 47
anaerobic bacteria
culture methods and media, 73–76
Hungate technique, 71–76
maintenance by cryopreservation, 76
maintenance by freeze-drying, 30–43, 76–80
cooling, 78
shelf life, 70, 79–80
drying, 169
subculturing, 76
viability, 79
cryopreservation
cooling, 48
glass beads, at −70°C, 45–50
over liquid nitrogen, 30, 62–63
on silica gel, 65–70
thawing, 49, 62, 106
drying
gelatin discs, 26
L-(liquid) drying, 59–60, 127–130
microdrying, 61–62

Bacteria (*continued*)
  paper strips or discs, 26
  pre-dried plugs, 26
  on silica gel, 66, 67, 70, 123–125
as food for protozoa, 199
freeze-drying, 30–43, 76–80
  for details *see* Freeze-drying, bacteria
gelatin discs, 51–55
on glass beads
  cooling, 48
  cryopreservation, 45–50
  thawing, precautions, 49
industrial and marine species, 57–63
  cooling, 62
  cryopreservation, 61–63
  freeze-drying, 57–59
  L-drying, 59–60
  microdrying, 61–62
  shelf life, cryopreservation, 61–63
  shelf life, drying, 59–60, 61–62
  suspension fluid, 58–59
  thawing, precautions, 62
L-(liquid) drying, 59–60, 127–130
microdrying, 61–62
reconstitution and rehydration, 35, 49, 54, 61, 62, 68, 79
shelf life
  by species, 39–43
  freeze-drying, 28, 37
  glass beads storage, 50
  industrial and marine species, 59–60, 61–63
  liquid paraffin, 122
strict anaerobes, Hungate technique, 71–76
survival, 39–43, 50, 54, 55, 61, 69, 70
viability, 35
see also Leptospira; Methanogenic bacteria; Phototropic bacteria
Bacteriophages
freeze-drying, 57–63
shelf life, 63
survival, 63
Balch serum tubes, 102
Beads, glass, drying of cultures, 125–127
Beads, glass, at −60°C to 76°C
  advantages, 45–46
  bacteria
    general principles, 29–30
    industrial and marine species, 62
    in liquid nitrogen, vapour phase, 62
    method, 46–49
    organisms successfully preserved, 50
    sensitive bacteria, 62
    shelf life, 50
    storage conditions, 49
  preparation, 46
Beads, porcelain, drying of cultures, 125–127

Belgium, National Collections, 9
BGP broth, 47, 75
BGPhlf broth, 75
Black light, fungi, induction of sporulation, 157–158
Blowtorches, 34, 78, 138, 173
Bovine albumin solution, 117
Brazil, culture collections, 19, 20
Bulgaria, National Collections, 9

Calcium chloride, as desiccant, 191
Camphor, 159
Capillary tubes
  cryopreservation of bacteria, 107–110
  preparation, 218–219
  storing and labelling, 224
Carnoy's fixative, 241
Centraalbureau voor Schimmelcultures, 13
Cerophyl leaf extract, 198
Cerophyl-Prescott (CP) agar, 200
Charcoal, activated, protection against photo-oxidation, 94, 96, 100, 127, 129
China, culture collections, 19
Chloroflexus growth medium, 87
Chromic acid, 218
Ciliates, media, 199
Clean rooms, 230
Cobalt chloride, in silica gel, toxicity, 123
Cold shock, 204
Coleccion Espanola de Cultivos Tipo, 14
Collectione Lieviti Industriali, 12
Cooked meat broth, 75
Cooling
  algae, 202–203, 210–211
  animal cells, 238
  bacteria
    on glass beads, 48
    industrial and marine, 62
    methanogenic, 107–109
    phototrophic, 91–92
  fungi, 68, 139
  plant tissue cultures, 253, 256, 257, 260–264
  protists, 201–205
  protozoa, 210–211, 221–224
  yeasts, 175–176
  general principles
    cold shock, 204
    effects of ice on cells, 210–211
    rapid cooling, 222
    slow cooling, 221–224
    two-stage, 210–211
    very rapid, 210–211
  see also Cryopreservation
Copper sulphate gelatin glue, 159
Cryomicroscopy, 204

Cryopreservation
  algae, 201–206
  animal cells, 236–239
    mouse embryos, 225
  bacteria, industrial and marine bacteria, 62–63
  cooling *see* Cooling
  cryoprotectants
    algae and protozoa, 211–212, 220–221
    animal cells, 236
    fungi, 141
    methanogenic bacteria, 105
    phototrophic bacteria, 91
    plant tissues, cultures, 256, 260
    yeasts, 176, 181–182
  cryoprotectants *for specific substances see*
    Cryoprotectants
  fungi
    spore suspensions, 140
    storage and survival, 153–157
  general principles, 29–30
    advantages, 66
    aerobic conditions, 90
    ulture stability, 266
    defined, 66
    and effect of very rapid cooling, 210–211
    glass beads, 45–50
    in vitro conservation in plant tissue cultures,
      266–267
    latent heat of fusion, 220
    predrying, 66
    on silica gel, 65–70
    storage on glass beads at −70°C, 29–30
    storage in liquid nitrogen, 30
    storage on silica gel, organisms successfully
      preserved, 68–70
  leptospira, 119
  methanogenic bacteria, 104–110
  phototrophic bacteria, 89–93
  plant tissue cultures, 250–265
    cell suspension, 253–255
    continuing problems and research
      challenges, 265–266
    equipment, 264–265
  protozoa, 201–206
    parasitic protozoa, 209–224
  rewarming, algae and protozoa, 212, 224–225
  yeasts, 65–70, 175–182
    stability of genetically modified strains, 180–
      182
  see also Beads, glass and porcelain; Silica gel
Cryoprotectants
  BGP broth, 47, 75
  dimethyl sulphoxide (DMSO), 91, 105, 176,
    202, 213–217, 221, 236, 256, 257, 259, 260,
    261, 262, 263, 264
  ethanol, 176

extension of cooling rate and viability, 181–182
  glucose, 176
  glycerol, 62, 91, 104, 105, 141, 176, 182, 202,
    211–212, 213–217, 220, 236, 256, 260
  hydroxyethyl starch (HES), 176, 216
  methanol, 176, 202
  polyethylene glycol (PEG), 176, 256, 259, 261
  sucrose, 176
Culture Collection of Algae, Norway, 13
Culture collections
  European collections, 8–18
  functions, 5–8
  major non-European collections, 19
Cultures, maintenance methods
  choice of, 21–24
  cryopreservation, 29–30
  drying, 25–27
  freeze-drying, 27–29
  subculture, 24–25
Cyanobacteria, freeze-drying, 207
Cysteine (hydrochloride), in media, 47, 72, 73, 75,
  104
Czechoslovakia, National Culture Collections, 9–
  10, 52, 53

Damage
  cold shock, 204
  latent heat of fusion, 220
  see also specific methods of maintenance and
    preservation
Desiccants
  calcium chloride, 191
  phosphorus pentoxide, 26, 27, 36, 53, 59, 68,
    78, 172–173
  disposal, 37
  silica gel, 26, 65–70, 123, 137, 146–150, 168–
    171
Deutsche Sammlung von Mikroorganismen und
  Zellkulturen (DSM), 12, 83, 104
para-Dichlorobenzene, mite deterrence, 159
Dimethyl sulphoxide (DMSO)
  as cryoprotectant, 91, 105, 176, 202, 213–217,
    221, 236, 256, 257, 259, 260, 261, 262, 263,
    264
  dilution factor, 92
Distilled water, storage of organisms, 122
DMSO *see* Dimethyl sulphoxide
Documentation, 223, 287–288
Drying
  cultures on glass or porcelain beads, 125–127
  gelatin discs, 26
  L-drying, 28, 59–60
    using simple apparatus, 127–130
  microdrying, 61–62
  paper strips and discs, 26

Drying (*continued*)
  phototrophic bacteria, 98
  predried plugs, storage of delicate bacteria, 26
  primary drying, 33
  principles, 25–27, 59–60, 127–130
  sand and soil, 26
  secondary drying, 34
  silica gel, 26, 65–70
  see also selected headings above, for further detail
Dulbecco's Modified Eagles Medium, 229

ECCO *see* European Culture Collections'
  Organization
Edwards high vacuum Modulyo freeze dryer,
  operation, 36–37
EM (Ellinghausen and McCullough) medium, 117
Escherichia coli, as food for protozoa, 199
Ethanol, as cryoprotectant, 176
Euglena gracilis medium, 188
European Collection of Animal Cell Cultures, 16
European culture collections, 8–18
European Culture Collections' Organization
  address, 20
  members' addresses, 9–18

Faecal extract, 75
Ferrous sulphide, 112
Filamentous fungi *see* Fungi, filamentous
Finland, National Collections, 9
Flagellates, media, 199
France, National Collections, 10
Freeze-drying
  anaerobic bacteria
    apparatus and equipment, 76–77
    opening ampoules, 79
    preparation of suspension, 77–78
    primary and secondary drying, 78
    survival and shelf life, 79–80
  bacteria
    ampoules, 33
    general notes, 27–29, 37–38, 80
    hazardous pathogens, 34
    organisms successfully preserved, 38–43
    pre-drying culture preparation, 32
    primary drying, 33
    secondary drying, 34
    storage conditions, 37
    survival (NCTC) before drying, after
      drying, and for various storage periods,
      39–43
    suspending fluids, 32
    viability counts, 35
  cyanobacteria, 207
  filamentous fungi, 138–139, 151

industrial and marine bacteria, 57–59
  methanogenic bacteria, 110–113
  phototrophic bacteria, 93–98
  protozoa, 206–207
  yeasts, 171–175
  principles, 27–29
    Hungate technique, 78
    operation of Edwards high vacuum Modulyo
      freeze dryer, 36–37
Freezing *see* cryopreservation
Fungi, filamentous
  age of culture for preservation, 135
  black light, induction of sporulation, 157–158
  cooling, 68, 139
  cryopreservation
    maintenance, 139–141
    on silica gel, 65–70
    storage, 152–157
    thawing, 140, 156
  drying, 137–138, 146–151
  freeze-drying 137–138, 151
  mite prevention, 158–159
  preparation of spore suspension, 138
  storage of organisms
    in distilled water, 122
    in liquid paraffin, 122
    shelf-life, 122, 137, 141, 142, 149, 150, 151,
      155
  subculturing, 135–137, 141–146

Gauze, use with ampoules, 33
Gelatin discs, maintenance of bacteria, 26, 51–55
Gelatin-suspending medium, 53
Gelrite, alternative to agar, 102
Gene banks, *in vitro* conservation of plant tissue
  cultures, 266–267
Germany, National Collections, 10–11
Germicidal solutions, 231
Glass ampoules *see* Ampoules
Glass capillary tubes *see* Capillary tubes
Glucose, in cryoprotectant mixtures, 176
Glycerol, as cryoprotectant, 62, 91, 104, 105, 141,
  176, 182, 202, 211–212, 213–217, 220, 236, 256,
  260
Gothenberg, University of Goeteborg Culture
  Collection, 14
Green sulphur bacteria
  characteristics, 82
  Pfennig's medium, 87
Guillard's medium, 189

Haemin solution, 74
Haemocytometer for viable cell counts, 234–236
Halophiles, 131

Handtorches *see* Blowtorches
Hazardous pathogens, 34
Heliozoans, media, 200
Hoechst 33258 stain, mycoplasmas, 241
Horse serum, 32, 58, 77, 172
Horse serum/sucrose/glucose medium, 80, 112, 172
Hungarian National Collection of Medical Bacteria, 12
Hungate technique, strict anaerobic bacteria, 71–76
    media, 73–75
Hybridomas, 229
Hydrogen, anaerobic jars, 72, 105, 113
Hydroxyethyl starch (HES), as cryoprotectant, 176, 216

IMI (International Mycological Institute), 16, 143
Industrial and marine bacteria *see* Bacteria, industrial and marine species
Information Center for European Culture Collections (ICECC), address, 7
Information centres
    European culture collections, 7–18
    regional and international, 7
    worldwide culture collections, 19–20
Ink, for labelling, 77, 138, 171
Inositol serum, bacterial suspensions, 32, 112, 138, 172
International Board for Plant Genetic Resources (IPBGR), *in vitro* gene banks, 267
Italy, Collectione Lieviti Industriali, 12

Japan, culture collections, 19
Jaworski's medium, 186

Kieselguhr, drying, and survival, 26
Korthof's medium, Leptospira, 116

L-(liquid) drying
    bacteria, 59–60
        industrial and marine bacteria, 59–60
    phototrophic bacteria, 98
    principle, 28
    survival, 130
    using simple apparatus, 127–130
Labelling
    ampoules, 219
    capillaries, 224
    type of ink, 77, 138, 171
Latent heat of fusion, cryopreservation, 220
Leptospira
    age of culture for preservation, 119
    cryopreservation, thawing, 119

EM (Ellinghausen and McCullough) medium, 117
    Korthof's medium, 116
    maintenance, 115–119
    reconstitution and rehydration, 119
    subculturing, 115–119
    thawing, 119
Light, black light, 157–158
Liquid nitrogen *see* Cryopreservation
Liquid-dried cultures *see* L-(liquid) drying
Liver extract, 75
Lyophilization *see* Freeze-drying

Malt-yeast extract agar (MY75S), 199
Marine bacteria *see* Bacteria, industrial and marine bacteria
Meat broth, 75
Media
    for amoebae, 199–200
    animal cells, 229
    ASP2 marine mineral medium, 188
    BGP broth, 47, 75
    BGPhlf broth, 75
    bovine albumin solution, 117
    Cerophyl-Prescott (CP) agar, 200
    Chloroflexus growth medium, 87
    for ciliates, 199
    Dulbecco's Modified Eagles Medium, 229
    EM (Ellinghausen and McCullough) medium, 117
    Euglena gracilis medium, 188
    faecal extract, 75
    ferrous sulphide, 112
    for flagellates, 199
    gelatin-suspending medium, 53
    Guillard's medium, 189
    haemin solution, 74
    for heliozoans, 200
    Hungate technique, 73–75
    inositol broth and serum, 32, 112, 138, 172
    Jaworski's medium, 186
    Korthof's medium, 116
    liver extract, 75
    malt-yeast extract agar (MY75S), 199
    meat broth, 75
    Mist. dessicans, 58
    MM10 broth, 73
    non-nutrient agar, 199
    Pfennig's medium, 87
    plant tissue cultures
        cryoprotectants, 256
        pre-growth, 256
        storage media, 244–245
    for protozoa, 197–201
    rabbit serum, inactivated, 116

Media (*continued*)
RBS-35, 218
Rhodospirillaceae growth medium, 84
Roswell Park Memorial Institute-1640, 229
rumen fluid, 74
S66 medium, 187
S88 marine mineral medium, 188–189
SES (soil extract medium with salts), 187
Sigma dehydrated cereal leaf preparation
C7171, 198
SL-10B trace element solution, 87
SL-12B trace element solution, 87
SL-6 trace element solution, 85
SNA (seawater nutrient agar), 190
Tetraselmis suecica medium, 190
Tween 80 solution, 117
VFA mixture, 74
VL broth, 75
VLhlf broth, 75
YM (yeast medium)
agar, 170
broth, 164
see also Suspension fluids
Methanogenic bacteria
age of culture for preservation, 104, 105
cooling, 107–109
cryopreservation, thawing, 106, 109–110
maintenance, 101–113
cryopreservation, 104–110
freeze-drying, 110–113
subculturing, 103–104
reconstitution and rehydration, 106, 113
shelf life, 103, 104, 106, 110–111
cryopreservation, 104, 106
freeze-drying, 110–111
thawing, precautions, 106, 109–110
viability, 106, 111, 113
see also Bacteria, anaerobic bacteria
Methanol, as cryoprotectant, 176, 202
Microbial Strain Data Network, address, 7
Milk, skimmed, suspension fluid, 95
Milk media, 126, 169
Mineral oil *see* Paraffin
Mist. dessicans, preparation, 58
Mites, damage to fungal cultures, 158–159
MM10 broth, 73
Monoclonal antibody-secreting hybridomas,
stability, 229
Most probable number (MPN) technique, 206
MSDN *see* Microbial Strain Data Network
Mycoplasmas
contamination of animal cells, 239–241
Hoechst 33258 stain, 241

National Collection of Agricultural Industrial
Microorganisms, 12

National Collection of Food Bacteria, 16
National Collection of Pathogenic Fungi, 16
National Collection of Plant Pathogenic Bacteria,
17
National Collection of Type Cultures, 17
National Collection of Wood Rotting Fungi, 17
National Collection of Yeast Cultures, 17
National Collections of Industrial and Marine
Bacteria, 16
NCTC *see* UK National Collection of Type
Cultures
Nitrogen, liquid
storage of bacteria, 30
see also Cryopreservation
Non-nutrient agar (NNA), 199
Norway, Culture Collection of Algae, 13

Paper replica method for yeasts, 168
Paper strips and discs, drying, survival, 26
Para-dichlorobenzene, 158–159
Paraffin, liquid
storage of bacteria and yeasts, 122
storage of fungi, 143–145
yeasts, 166–167
Pfennig's medium
green sulphur bacteria, 87
purple sulphur bacteria, 85
Phosphorus pentoxide
disposal, 37
use in freeze-drying, 26, 27, 36, 53, 59, 68, 78,
172–173
Photo-oxidation, use of activated charcoal, 94, 96,
100, 127, 129
Phototrophic bacteria
age of culture for preservation, 91
cooling, 91–92
cryopreservation
cooling, 92
thawing, 92
maintenance, 81–100
cryopreservation, 89–93
culture methods and media, 83–88
freeze-drying
aerobic conditions, 93–95
anaerobic conditions, 96–98
general notes, 98–100
liquid-drying, 98
subculturing, 88–89
purple, green, and *incertae sedis*, 82
reconstitution and rehydration, 92, 95, 97
shelf life
freeze-drying, 95, 98
L-drying, 98
survival, 92, 96
thawing, precautions, 92
viability, 92, 98

Plant Genetic Resources, International Board, *in vitro* gene banks, 267
Plant pathogens, storage in distilled water, 122
Plant tissue cultures
 applications, 244
 callus cultures, 248
 conservation, 266
 cooling, 253, 256, 257, 260–264
 cryopreservation, 250–265
  callus, 257–259
  cell suspension cultures, 253–255
  equipment, 264–265
  principles, 250–253
  protoplasts, 255–257
  shoot cultures, 259–262
  somatic embryos, 262–264
  thawing, 253–255, 256–259, 260–264
 culture maintenance
  normal growth storage, 246
  slow growth storage, 247, 248–250
  suspension of growth, 248
 culture storage, 244–245
 multiple modes of storage, 245
 pollen, 248
 reconstitution and rehydration, 252, 255, 256, 259
 shoot cultures, 249–250
 somaclonal variation, 247
 somatic embryos, 262
 systems and applications, 243–244
 thawing, precautions, 253–255, 256–259, 260–264
Plugs, predried, storage of delicate bacteria, 26
Polish Collection of Microorganisms, 13
Polyethylene glycol (PEG), as cryoprotectant, 176, 256, 259, 261
Polysilicate plates, alternative to agar, 102
Porcelain beads *see* Beads
Porphyrines, protection against photo-oxidation, 94
Portuguese Yeast Culture Collection, 13
Protozoa
 age of culture for preservation, 191, 202
 air-drying, 206
 amoebae, 199
 axenic media, 185, 186, 191, 193, 194, 198
 ciliates, 199
 cooling, 201–205
  effects, 210–211
 cryopreservation, 201–206
  by species, thawing, 212–217
  'cold-shock', 204
  general principles, thawing, 206, 224–225
 culture observations, 195–196
 culture vessels, 196
 dark ground illumination microscopy, 196
 encystment, 199, 206

 flagellates, media, 199
 freeze-drying, 206–207
 heliozoans, 200
 media, 197–199
 shelf life, 198
 stabilisation temperature, 201
 storage, 198, 201
 subculturing, 195–201
 survival, 204
 viability, 201, 205–206
Protozoa, parasitic, 225
 cooling, 221–224
 cryopreservation, 209–224
 rapid cooling, 222
 rewarming, 222–223
 shelf life, 212
  cryopreservation, 225
 storage and documentation, 223–224
 survival assays algae and protozoa, 225–226
 thawing, precautions
  by species, 212–217
  general principles, 224–225
Purple sulphur bacteria
 characteristics, 82, 85
 Pfennig's medium, 85

Rabbit serum, inactivated, 116
Raffinose, for lyophilization, 94, 95
RBS-35, 218
Recombinant clones and plasmids, cryopreservation, 69, 180–181
Reconstitution and rehydration
 algae, 206
 animal cells, 238
 bacteria, 35, 49, 54, 61, 62, 68, 79
 general principles
  distilled water, 123
  L-drying, 130
  liquid paraffin, 122
  silica gel, 127
 leptospira, 119
 methanogenic bacteria, 106, 113
 phototrophic bacteria, 92, 95, 97
 plant tissue cultures, 252, 255, 256, 259
 yeasts, 169, 170, 173, 176, 179
Record keeping, and documentation, 287–288
references, 269–285
Rehydration
 activated charcoal, 131
 phototrophs, 97
Resazurin solution, 74
Resuscitation *see* Reconstitution and rehydration
Revival *see* Reconstitution and rehydration;
 Survival
Rhodospirillaceae growth medium, 84

Roswell Park Memorial Institute-1640, 229
Rumen fluid, 74

S66 medium, 187
S88 marine mineral medium, 188–189
'Safe-deposit' facilities, 6
Safety precautions
    ampoules, 225
    hazardous pathogens, 34
    liquid nitrogen, 99
Seawater nutrient agar (SNA), 190
Serial transfers, 134
Service collections
    functions, 5–20
    information centres, 7
    obtaining cultures, 5
    patent procedures, 6–8
    recording, 5–6
    'safe-deposit' facilities, 6
    selection of maintenance methods, 5
    services provided, 6–8
Shelf life
    algae
        drying, 206
        freeze-drying, 207
    bacteria
        anaerobic bacteria, 79–80
            Hungate method, 70
        by species, 39–43
        freeze-drying, 28, 37
        glass beads storage, 50
        industrial and marine, 60, 61–63
        methanogenic, 103, 104, 106, 110–111
        phototrophic, 89, 95, 98
    bacteriophages, 63
    cryopreservation
        general principles, 30
        industrial and marine bacteria, 61–63
        leptospira, 119
        methanogenic bacteria, 104, 106
        parasitic protozoa, 225
        yeasts, 177, 180
    drying
        algae, 206
        anaerobic bacteria, 79–80
        general principles, 26
        yeasts, 169
    freeze-drying
        algae, 207
        anaerobic bacteria, 79–80
        bacteria, 28, 37
        methanogenic bacteria, 110–111
        phototrophic bacteria, 95, 98
        yeasts, 174

fungi
    at growth temperature, 141
    cryopreservation, 155–156
    drying, 149, 150
    freeze-drying, 151
    refrigerated, 142
    subculturing, 137
    under liquid paraffin, 122, 143
    under oil, 122, 143
    in water, 146
glass beads storage of bacteria, 50
L-drying, 130
    general principles, 28
    industrial and marine bacteria, 60
    phototrophic bacteria, 98
leptospira, 119
liquid paraffin
    bacteria, 122
    fungi, 122
microbes that do not survive well, 131–132
parasitic protozoa, 212
silica gel, yeasts, 170
subculturing
    general principles, 24–25
    leptospira, 119
    methanogenic bacteria, 103
    phototrophic bacteria, 89
    yeasts
        agar slants, 166
        broth, 165
        in water, 168
yeasts
    cryopreservation, 177, 180
    drying, 169
    freeze-drying, 174
    silica gel, 170
    in straws, 180
    subculturing, 165, 166, 168
Sigma dehydrated cereal leaf preparation C7171, 198
Silica gel
    checking cultures for viability, 137
    cryopreservation of bacteria
        advantages, 70
        details of method, 65–68
        organisms successfully preserved, 68–69
    dispensing and drying fungal spore suspension, 137
    drying procedure, 66, 67, 70, 123–125
        survival, 26
    fungi, storage and survival, 146–150
    preparation, 137
    toxicity of cobalt chloride, 123
    yeasts, 168–171
SL-12B trace element solution, 87
SL-6 trace element solution, 85

SNA (seawater nutrient agar), 190
Sodium glutamate, in suspension fluids, 58
Sodium sulphide solution, phototrophic bacteria,
    85
Soil extract medium with salts, 187
Soil/sand
    dispensing fungal spore suspension, 138
    drying, survival, 26
    storage and survival of fungi, 150–151
Sorbitol, as cryoprotectant, 260, 263
Spain, Coleccion Espanola de Cultivos Tipo, 14
Stability, 22, 30, 54, 69, 98, 146, 162, 163, 166,
    170, 171, 174, 181–2
Straws, cryopreservation of yeasts, 177–180
Subculturing
    algae, 185–195
    animal cells, 232–236
        adherent cell lines, 232–233
    bacteria
        anaerobic bacteria, 76
        methanogenic bacteria, 103–104
        phototrophic bacteria, 88–89
    filamentous fungi, 135–137, 141–146
    general principles, 24–25
    leptospira, 115–119, 119
    plant tissue cultures, 244–250
    protozoa, 195–201
    yeasts, 163–168
Sucrose, as cryoprotectant, 176
Sulphate-reducing bacteria, 58
suppliers, list, 289–292
Survival
    bacteria, 39–43, 50, 54, 55, 61, 69, 70
    bacteriophages, 63
    general principles, 131–132
    liquid-dried cultures, 130
    phototrophic bacteria, 92, 96
    and revival
        after drying, 127
        storage under liquid paraffin, 122
        storage in water, 123
    see also Reconstitution; Shelf life; Viability;
        specific organisms
Suspension fluids
    with activated charcoal, 94, 96, 100
    algae, 191
    bacteria
        anaerobic bacteria
            Hungate technique, 73–75, 77–78
            McCartney bottles, 75, 78
        cryopreservation on silica gel, 67
        for freeze-drying, 32
        industrial and marine bacteria, *mist
            desiccans*, 58
        maintenance on gelatin disks, 53
        maintenance on glass beads, 47

methanogenic bacteria, 110, 112
dimethyl sulphoxide (DMSO), 105
fungi, 67, 135, 139
for glass beads, 47
horse serum, 32, 58, 77, 172
horse serum/sucrose/glucose medium, 80, 112,
    172
industrial and marine bacteria, 58–59
inositol
    broth, 32
    serum, 32
inositol serum, 32, 112, 138, 172
protozoa, 195 196
skimmed milk, 95
yeasts, 172
see also Media
Sweden, University og Goeteborg Culture
    Collection, 14
Switzerland, Centre de Collection de Type
    Microbien, 14

Tetraselmis suecica medium, 190
Thawing, rate of thawing, 109
Thawing, precautions
    algae, 202–203, 206
    animal cells, 238
    bacteria
        on glass beads, 49
        industrial and marine, 62
        methanogenic, 106, 109–110
        phototrophic, 92
    fungi, 140, 156
    leptospira, 119
    parasitic protozoa
        by species, 212–217
        general principles, 224–225
    plant tissue cultures, 253–255, 256–259, 260–
        264
    protists, 206
    yeasts, 173, 176
Tripticane, 73
Trypanosomiasis, WHO World Collaborating
    Centre, 221
Turkey, National Collection, 15
Tween 80 solution, 117

UK National Collection of Industrial and Marine
    Bacteria, 57
UK National Collection of Type Cultures, 32
UK National Collection of Yeast Cultures, 163
    see also National Collections
Uppsala University Culture Collection of Fungi, 14
USA, National Culture Collections, 19
USSR, National Culture Collections, 20, 65–66

VFA mixture, 74
Viability
  algae, 205
  anaerobic bacteria, 79
  animal cells, 234
  bacteria, 35
  general principles, 22, 131
  methanogenic bacteria, 106, 111, 113
  parasitic protozoa, 225
  phototrophic bacteria, 92, 98
  protozoa, 206
  simple method for measurement, 130, 131
  yeasts, 164, 173
VL broth, 75
VLhlf broth, 75

World Data Center for Collections of Cultures of
  Microorganisms (WDC), address, 7
World Federation for Culture Collections, address,
  20

Yeasts
  on agar slants, 166, 167
  age of culture for preservation, 162, 168, 170,
    172, 175
  available methods, 163
  in broth, 165
  comparison of preservation methods, 164
  cooling, 175–176
  cryopreservation
    genetically modified strains, 180–182

liquid nitrogen, 175–180
on silica gel, 65–70
in straws, 177–180
thawing, 173, 176
cryoprotectants, 176, 181–182
drying
  paper replica method, 168
  silica gel method, 168–171
freeze-drying, 171–175
recombinant clones and plasmids, 180–181
reconstitution and rehydration, 169, 170, 173,
  176, 179
rehydration, 169, 170
revival, 173, 176
shelf life
  cryopreservation, 164, 177, 180
  drying, 169
  freeze-drying, 174
  silica gel, 164, 170
with special requirements, 166, 167
stability of genetically modified strains, 180–182
storage
  in distilled water, 122
  in liquid paraffin, 122
  in straws, 180
subculturing, 163–168
  agar with oil overlay, 166
  agar slants, 165–166
  broth, 164–165
  in water, 168
survival, factors affecting, 161, 170, 180
thawing, precautions, 173, 176
viability estimation, 164, 173